Construction industry questions and answers: a practical approach

Pat Perry

Thomas Telford

Published by Thomas Telford Publishing, Thomas Telford Ltd, 1 Heron Quay, London E14 4JD.
URL: http://www.thomastelford.com

Distributors for Thomas Telford books are
USA: ASCE Press, 1801 Alexander Bell Drive, Reston, VA 20191-4400, USA
Japan: Maruzen Co. Ltd, Book Department, 3–10 Nihonbashi 2-chome, Chuo-ku, Tokyo 103
Australia: DA Books and Journals, 648 Whitehorse Road, Mitcham 3132, Victoria

First published 2003

Also in this series from Thomas Telford Books
Health and safety: questions and answers. Pat Perry. ISBN 07277 3240 4
Fire safety: questions and answers. Pat Perry. ISBN 0 7277 3239 0
Risk assessment: questions and answers. Pat Perry. ISBN 0 7277 3238 2
CDM questions and answers: a practical approach 2nd edition. Pat Perry. ISBN 0 7277 3107 6

A catalogue record for this book is available from the British Library

ISBN: 0 7277 3233 1

© Pat Perry and Thomas Telford Ltd, 2003

Any safety sign or symbol used in this book is for illustrative purposes only and does not necessarily imply that the sign or symbol used meets any legal requirements or good practice guides. Before producing any sign or symbol, the reader is recommended to check with the relevant British Standard or the Health and Safety (Safety Signs and Signals) Regulations 1996.

Throughout the book the personal pronouns 'he', 'his', etc. are used when referring to 'the Client', 'the Designer', 'the Planning Supervisor', etc., for reasons of readability. Clearly, it is quite possible these hypothetical characters may be female in 'real-life' situations, so readers should consider these pronouns to be grammatically neuter in gender, rather than masculine.

Typeset by Alex Lazarou, Surbiton, Surrey
Printed and bound in Great Britain by MPG Books, Bodmin, Cornwall

Biography

Pat Perry, MCIEH, MIOSH, FRSH, MIIRM, qualified as an Environmental Health Officer in 1978 and spent the first years of her career in local government enforcing environmental health laws, in particular health and safety law, which became her passion. She has extensive knowledge of her subject and has served on various working parties on both health and safety and food safety. Pat contributes regularly to professional journals, e.g. *Facilities Business,* and has been commissioned by Thomas Telford Publishing to write a series of health and safety books.

After a period in the private sector, Pat set up her own environmental health consultancy, Perry Scott Nash Associates Ltd, in the latter part of 1988, and fulfilled her vision of a 'one-stop shop' for the provision of consultancy services to the commercial and retail sectors.

The consultancy has grown considerably over the years and provides consultancy advice to a wide range of clients in a variety of market sectors. Leisure and retail have become the consultancy's major expertise and the role of planning supervisor and environmental health consultant is provided on projects ranging from a few hundred thousand pounds to many millions, e.g. new public house developments and major department store refits and refurbishments.

Perry Scott Nash Associates Ltd have strong links to the enforcing agencies; consultants having come mostly from similar backgrounds and approach projects and all the issues and concerns associated with legal compliance with pragmatism and commercial understanding.

Should you wish to contact Pat Perry about any issue in this book, or to enquire further about the consultancy services offered by Perry Scott Nash Associates Ltd, please contact us direct at:

Perry Scott Nash Associates Ltd
Perry Scott Nash House
Primett Road
Stevenage
Herts
SG1 3EE

Alternatively phone, fax or email on:

Tel: 01438 745771
Fax: 01438 745772
Email: p.perry@perryscottnash.co.uk

We would also recommend that you visit our website at:
www.perryscottnash.co.uk

Acknowledgements

My sincere thanks go to Maureen for her never ending support and encouragement and to Janine and the Business Support team at Perry Scott Nash Associates Ltd for typing all the handwritten manuscripts with such patience and efficiency.

Author's note

Many of the publications referenced in this book are available for download on a number of websites, e.g.:

- www.hse.gov.uk/pubns/index.htm
- www.hsebooks.co.uk

Also, guidance on the availability of books is available from the HSE Info Line on 0541 545 500.

Contents

Appendices

1

Introduction

Why is construction safety important?

More people die from injuries and ill-health as a result of working on construction projects than in any other working environment.

Construction works, and all the ancillary activities, are hazardous with the potential to cause deaths, severe injuries and long-term ill-health.

All employers and others have duties to ensure the health, safety and welfare of their employees and also of those who are not in their employ.

No individual should expect to go to work and be at an undue risk of being killed, injured or exposed to long-term occupational ill-health.

What are some of the construction industry's statistics?

The Health and Safety Executive (HSE) collate accident and ill-health statistics from the accident notifications they receive. From these statistics they can confirm that the construction industry has the worst health and safety record of any employment sector.

Statistics for the year 2000/2001 show the following:

Fatal injuries by accident kind, all workers	Percentage
All falls from height	44
Struck by moving vehicle	17
Struck by moving or falling object	8
Trapped by something collapsing or overturning	17
Other	14

Total fatalities in 2000/2001 in the construction sector were 106.

Despite increased awareness of health and safety standards, the number of fatalities increased in those years from a total of 81, which was the number of fatalities recorded in the years 1999/2000.

While a significant number of individuals die from their injuries on construction sites, a considerably higher number of individuals suffer a major injury which could disable them for life.

During the period 2000/2001, 4268 people suffered a major injury while working in construction.

These statistics are as shown below:

Number of construction workers with major injuries by cause	Percentage
Falls from height	37
Slip, trip or fall on the same level	21
Struck by moving vehicle	2
Struck by moving or falling object	18
Injured while handling, lifting or carrying	8
Other	14

What can employers do about the causes of accidents and injuries in the construction industry?

Employers need to ensure that they comply with health and safety legislation and provide safe working conditions and practices for their employees.

By understanding some of the most hazardous activities, employers can ensure that they introduce safe systems of work to reduce the possibility of adverse consequences for those activities.

Falls from height are some of the commonest causes of death and serious injury. The Health and Safety Executive have introduced robust inspection procedures and checks on all activities involving working at heights. Falls from height are generally preventable, provided safe systems of work are followed and suitable protective equipment used.

Surely accidents will always happen because construction workers want to be 'macho' and will never follow safe systems of work, etc.?

The construction industry suffers from a poor reputation in respect of its 'macho' culture and construction workers often expect to take risks, fail to follow procedures or refuse to wear protective clothing or equipment.

However, attitudes can change and research from the Health and Safety Executive has shown that management has a huge role to play in changing the culture and attitude of workers to safety.

The Health and Safety Executive found that the positive attitude of management could have prevented approximately 70% of fatalities. Positive attitude means listening to the concerns of individuals and taking steps to reduce hazards and risks by either changing the way the job is done, providing the proper equipment or insisting that personal protective equipment is worn or used.

Is anyone else affected by the hazards and risks associated with the construction industry?

Construction workers are by far the most affected by the hazards and risks of construction works and, on average, two people die every week because they work on a construction site.

But they are not the only people to be affected. Because of the poor safety management of so many construction sites and the associated activities, at least one member of the public is killed every week, and several could be seriously injured.

Members of the public are killed by moving vehicles, falling objects, poorly erected scaffolding and collapsing buildings.

Is there evidence to show whether any particular trade is more at risk of injury than others?

The Health and Safety Executive have collated statistics which show that the following trades are more susceptible to death or injury:

Labourer
Roofer
Electrician
Demolition Contractor
Carpenter
Painter

Highest

Lowest

The Health and Safety Executive will expect employers to pay particular attention to the safe systems of work which are required to be followed by the most vulnerable trades.

Is it only fatalities and major injuries which are the concerns regarding construction safety?

There are two other major concerns with respect to construction safety:

- 'over three day' injuries
- occupational ill-health.

'Over three day' injuries are required to be reported to the Health and Safety Executive by employers or the self-employed.

Over three day injuries are those in which an employee will be off work for more than three days as a result of an accident at work. The injury is not severe enough to cause a 'major' injury — usually a broken limb, loss of sight, unconsciousness, hospitalisation, etc. — but is sufficient to cause the operative to be unable to attend work.

Types of injury which fall into this category will be:

- manual handling back injuries
- strained muscles
- slips and trips
- cuts and bruising
- twisted ankles and/or knees
- head injuries.

During the years 2000/01 there were over 9000 over three day injuries reported to the Health and Safety Executive. By far the greatest cause were manual handling injuries.

Occupational ill-health is also of major concern in the construction industry and includes illnesses which may be developed due to exposure to small quantities of hazard over prolonged periods of time.

The number of people suffering occupational ill-health is almost impossible to determine as many people will leave the industry without reporting their disease.

Types of occupational ill-health will be:

- lung disease from breathing in dust
- asbestosis
- industrial deafness
- vibration white finger
- severe industrial dermatitis.

Severe conditions of any of the above, plus any of a vast number of other industrial diseases, will result in the sufferer being unable to work. Not only that, usually the quality of life is severely affected and the individual is often disabled and unable to lead an active life.

Many incidences of occupational ill-health are preventable with good safety management controls and adherence to the Control of Substances Hazardous to Health Regulations.

2

Legal overview

What legislation is applicable to construction projects?

The following legislation can be applied to construction projects:

- Health and Safety at Work Etc. Act 1974
- Health and Safety (First Aid) Regulations 1981
- Electricity at Work Regulations 1989
- Noise at Work Regulations 1989
- Construction (Head Protection) Regulations 1989
- Personal Protective Equipment at Work Regulations 1992
- Manual Handling Operations Regulations 1992
- Reporting of Injuries, Diseases and Dangerous Occurrences Regulations 1995
- The Construction (Health, Safety and Welfare) Regulations 1996
- The Health and Safety (Safety Signs and Signals) Regulations 1996
- The Confined Spaces Regulations 1997
- The Provision and Use of Work Equipment Regulations 1998
- The Lifting Operations and Lifting Equipment Regulations 1998
- The Management of Health and Safety at Work Regulations 1999

- The Fire Precautions (Workplace) Regulations 1997 and 1999
- The Pressure Systems Safety Regulations 2002
- Control of Asbestos at Work Regulations 2002
- Control of Lead at Work Regulations 2002
- Control of Substances Hazardous to Health Regulations 2002
- The Dangerous Substances and Explosive Atmosphere Regulations 2002.

In addition to the above list, one of the key pieces of legislation for construction safety is:

- The Construction (Design and Management) Regulations 1994.

What are the key things I need to know about the Health and Safety at Work Etc. Act 1974?

The Health and Safety at Work Etc. Act 1974 is the 'underpinning' legislation which governs virtually all other health and safety law.

The Act sets out the general parameters of what is expected of employers and other persons in respect of ensuring their health, safety and welfare.

Regulations are subsidiary legislation made under the enabling powers of the Health and Safety at Work Etc. Act 1974. Contravening Regulations is an offence and prosecutions can be brought regarding each offence. In addition, there may be a breach of the more general principles of health and safety enshrined in the 1974 Act and additional charges could be brought under various sections.

The main sections of the 1974 Act are as follows.

Section 2 — General duty of employers to their employees

It shall be the duty of every employer to ensure, so far as is reasonably practicable, the health, safety and welfare at work of all his employees ... the matters to which that duty extends include:

1. Provision and maintenance of plant and equipment, and systems of work that are safe and without risks to health.
2. Arrangements for ensuring safety and absence of risks to health in connection with the use, handling, storage and transport of articles and substances.
3. The provision of such information, instruction and training as is necessary to ensure the health and safety at work of his employees.
4. The maintenance of any place of work under the employer's control in a condition which is safe and without risk to health and the provision of means of access and egress that are safe.
5. The provision and maintenance of a working environment that is safe, without risks to health and adequate with regard to facilities and arrangements for their welfare at work.
6. Provision of a written statement of his policy in respect of health and safety of his employees.

Section 3

It shall be the duty of every employer, to conduct his undertaking in such a way as to ensure, so far as is reasonably practicable, that persons not in his employment are not exposed to risks to their health and safety.

Section 4

It shall be the duty of each person who has control of premises to take such measures to ensure that premises, means of access to and egress from available for use by persons using the premises are safe and without risks to health.

Section 7

It shall be the duty of every employee while at work:

1. To take reasonable care for the health and safety of himself and of other persons who may be affected by his acts or omissions at work; and
2. To co-operate with his employer so far as is necessary so as to enable the employer to undertake statutory duties.

OFFENCES, PENALTIES AND PROSECUTIONS

1. Offences

These include the following:

(a) Failing to comply with the general duties on employers, employees, the self-employed, persons in control of premises, manufacturers, etc.
(b) Failing to comply with any requirement imposed by Regulations made under the Act.
(c) Obstructing or failing to comply with any requirements imposed by inspectors in the exercise of their powers.
(d) Failing to comply with an Improvement or Prohibition Notice.
(e) Failing to supply information as required by a Notice served by the Health and Safety Commission (e.g. investigations into major accidents).
(f) Failing to comply with a Court Order to remedy the cause of an offence.

2. Penalties

Most offences under the Health and Safety law are 'triable either way', i.e. summarily or on indictment.

Breaches of employers' duties under Section 2 of the Health and Safety at Work Etc. Act 1974 carry a maximum fine of £20 000 *per offence* if tried summarily.

Breaches of employers' duties under the numerous Regulations enacted under the Health and Safety at Work Etc. Act 1974 carry fines of up to £5000 per offence if tried summarily. In some circumstances, fines are up to £20 000 per offence.

Where the case is heard in the Crown Court, fines are unlimited.

Breaches of Improvement or Prohibition Notices carry either six months' imprisonment if heard in the Magistrates' Court or up to two years' imprisonment if heard at Crown Court.

3. Prosecutions

(a) Offences by Companies, Corporate Bodies and Directors (Health and Safety at Work Etc. Act 1974, Section 37)

The Health and Safety statutes place duties upon limited companies and/or functional directors.

Where an offence is committed by a body corporate, senior persons in the hierarchy of the Company may be *individually liable*.

If the offence was committed with the consent or connivance of, or was attributable to any neglect on the part of the following persons, that person himself is guilty of an offence and liable to be prosecuted:

- any functional director
- manager
- Secretary (Company)
- other similar officer of the company
- anyone purporting to act as the above.

The conditions for liability under Section 37 are:

- Did the person act as the Company?
- If he acted in that capacity, did he act with neglect?

Directors, Managers and Company Secretaries are personally liable for ensuring that Corporate Safety duties are performed throughout the Company. They may be able to delegate the specific responsibilities but that does not absolve them of liability.

(b) Offences due to the act of 'another person'

Section 36 of the Health and Safety at Work Etc. Act 1974 states that where an offence is due to the act or default of another person, then:

- that other person is guilty of the offence, and
- a second person (e.g. body corporate) can be charged and convicted whether or not proceedings are taken against the first-mentioned person.

What are the key things I need to know about the Management of Health and Safety at Work Regulations 1999?

Employers must make suitable and sufficient assessment of the risks to health and safety of their employees and to non-employees affected by their work. Each risk assessment must identify the measure necessary to comply with relevant statutory provisions.

Risk assessments must be in writing where there are more than five employees and they must be reviewed regularly.

Employers must introduce appropriate arrangements for effective planning, organisation, control, monitoring and review of the preventative and protective measures. These arrangements must be in writing where there are five or more employees.

Where appropriate, employees must have health surveillance.

Employers must establish and effect appropriate procedures to deal with emergencies, e.g. evacuation, major chemical spillage, explosion, etc. Employees must stop work immediately and proceed to a place of safety if exposed to serious imminent and unavoidable danger.

Employers must provide comprehensive and understandable information to all employees on risks identified in the risk assessments, emergency procedures, preventative and protective measures, competent personnel.

Employees must receive appropriate health and safety training on recruitment and throughout their employment.

Employers must appoint a 'competent person' to assist in undertaking the measures necessary to comply with statutory provisions. Competent persons can be employees or external advisers/consultants are permissible if there are no suitable employees.

Employers must consider the health and safety of young persons at work, pregnant women and nursing mothers.

Employers must carry out Fire Risk Assessments.

Contact must be made by employers with external services, e.g. fire, police, emergency services, so that any necessary measure can be taken in the event of an emergency or rescue.

Multi-occupied sites must have a plan for co-operation and co-ordination with one employer taking the lead role in respect of health and safety.

Employers must take into account the capabilities and training of all employees before assigning them tasks, etc.

Employees are under a duty to use equipment, materials, etc. provided to them by their employer in accordance with safe systems of work, any training, etc.

Employees must inform their employer of any matter relating to their own or others' health and safety.

Temporary workers are to be afforded health and safety protection, information on hazards, risks, health surveillance as appropriate.

What are the key things I need to know about the Health and Safety (First Aid) Regulations 1981?

Every employer must provide equipment and facilities which are adequate and appropriate in the circumstances for administering first aid to employees.

Employers must make an assessment to determine the needs of the workplace. First aid precautions will depend on the type of work, and therefore the risk, being carried out. Employers should consider the need for first aid rooms, employees working away from the premises, employees of more than one employer working together, non-employees. Once an assessment is made, the employer can work out the number of first aid kits necessary by referring to the Approved Code of Practice.

Employers must ensure that adequate numbers of 'suitable persons' are provided to administer first aid. A 'suitable person' is someone trained in first aid to an appropriate standard. In appropriate circumstances the employer can appoint an 'appointed person' instead of a first aider. This person will take charge of any situation, e.g. call an ambulance, and should be able to administer emergency first aid.

Employers must inform all employees of their first aid arrangements and identify trained personnel.

What are the key things I need to know about the Noise at Work Regulations 1989?

Employers must carry out Noise Assessments where employees are likely to be exposed to noise of:

- 'first action level' or above 85 dB(A)
- 'peak action level' or above 200 pascals.

Competent persons must carry out the assessments and identify employees exposed to the noise.

Noise Assessments must be reviewed.

Records must be kept of all Noise Assessments.

Employers have a duty to reduce the noise to which employees are exposed, if it measures 90 dB(A) or above.

Ear protection zones must be designated where appropriate.

Information, instruction and training must be given to all employees.

What are the key things I need to know about the Electricity at Work Regulations 1989?

All systems, plant and equipment must be designed to ensure maximum practical level of safety.

Installation and maintenance must reflect specific safety requirements.

Access, light and working space must be adequate.

Means of cutting off power and isolating equipment must be available.

Precautions must be taken to prevent charging.

No live working should be permitted unless absolutely essential.

Specific precautions are to be taken where live working is essential.

All persons must be effectively trained and supervised.

Responsibility for observing safety policy must be clearly defined.

All equipment and tools must be appropriate for safe working.

All personnel working on electrical systems must be technically competent and have sufficient experience.

Work activity shall be carried out so as not to give rise to danger.

Electrical systems must be constructed and maintained to prevent danger. 'Danger' means the risk of injury.

What are the key things I need to know about the Manual Handling Operations Regulations 1992?

Employers must, as far as is reasonably practicable, avoid the need for employees to undertake manual handling operations at work which involve risk of injury.

Employers must carry out risk assessments of all manual handling operations which cannot be avoided. Assessments must be in writing.

Employers must take appropriate steps, following assessments, to reduce the risk of injury to the lowest level reasonably practicable.

Employers must take appropriate steps to provide employees who are undertaking manual handling operations with general indications and/or precise information on the weight, etc. of each load and handling activity undertaken.

Employers must review assessments regularly, especially if there is any cause to believe that there has been a significant change in handling operation.

Employees must make full and proper use of any system of work provided by the employer concerning steps to reduce risks.

Consideration must be given to repetitive actions involving small weights.

What are the key things I need to know about the Personal Protective Equipment Regulations 1992?

Employers must ensure that suitable personal protective equipment (PPE) is provided to employees who may be exposed to risks to their health and safety, except where the risk has been controlled adequately by other means.

PPE must be suitable and appropriate to the risks and workplace conditions. It must suit the worker due to wear it and afford adequate protection.

If more than one sort of PPE must be worn, they shall be compatible.

PPE must be assessed to ensure that it is suitable for the tasks. Any assessments shall be reviewed regularly.

PPE must be maintained in an efficient state, in efficient working order and in good repair.

Where PPE is needed, suitable accommodation for it must be provided.

Adequate and appropriate instruction and training must be given to employees in the use of PPE.

Employees must use PPE provided, return it to its accommodation after use and report loss or defects.

Employers must consider PPE as a 'last resort' and must address all other risk reduction methods.

What are the key things I need to know about the Reporting of Injuries, Diseases and Dangerous Occurrences Regulations 1995?

Where any person dies or suffers any of the injuries or conditions specified in Appendix 1 of the Regulations, or where there is a 'dangerous occurrence', as specified in Appendix 2 of the Regulations as a result of work activities, the 'responsible person' must notify the relevant enforcing authority.

Notification must be by telephone or fax and confirmed in writing within seven days.

Where any person suffers an injury not specified in the Appendix but which results in an absence from work of more than three calendar days, the 'responsible person' must notify the enforcing authority in writing.

The 'responsible person' may be the employer, the self-employed, someone in control of the premises where work is carried out or someone who provided training for employment.

Where death of any person results within one year of any notifiable work accident, the employer must inform the relevant enforcing authority.

When reporting injuries, diseases or dangerous occurrences the approved forms — either F2508 or F2508A — must be used.

Records of all injuries, diseases and occurrences which require reporting must be kept for at least three years from the date they were made.

Accidents to members of the public which result in them being taken to hospital as a result of the work activity must be reported.

Incidents of violence to employees which result in injury or absence from work must be reported.

What are the key things I need to know about the Lifting Operations and Lifting Equipment Regulations 1998?

Lifting equipment and operations are covered by the Regulations, including lifts, hoists, eye bolts, chains, slings, etc.

Lifting equipment must be adequate in strength and stability for each load.

All equipment used for lifting people must be safe so that they cannot be crushed, trapped, struck by or fall from the lifting carrier.

Where safety ropes and chains are used they must have a safety co-efficient of at least twice that required for general lifting operations.

Lifting equipment must be installed or positioned in such a way that it reduces the risk of the equipment striking a person or of the load drifting, falling freely or being released unintentionally.

Suitable devices must be available to prevent persons from falling down a shaft or hoist-way.

Equipment used for lifting, including that used for lifting people, must be marked with safe working loads.

If equipment is not suitable for lifting people it must be marked accordingly.

Lifting operations must be properly planned, appropriately supervised and carried out in a safe manner.

Lifting equipment must be regularly inspected and tested and a report of its condition produced:

- before being used for the first time
- after assembly and being put into service for the first time at a new site.

Lifting equipment used for lifting *people* must be examined and tested every *six months*.

All other lifting equipment, e.g. goods hoists, must be examined and tested every *twelve months*, unless the competent person deems them to need more frequent inspections.

Records must be kept, although these can be electronic.

Persons that are carrying out examination and testing must be competent.

What are the key things I need to know about the Provision and Use of Work Equipment Regulations 1998?

Employers must ensure that work equipment is suitable by design, construction or adaptation for the work for which it is provided.

Employers must take into account any risks in the location where the equipment will be used, e.g. wet areas and electrical equipment.

Work equipment must be maintained in a suitable condition and in good working repair. Where there is a maintenance log, it must be kept up to date.

Equipment will have to be guarded where necessary, be able to be isolated from power sources, be used in the right environment, have display warnings, etc.

Employees must be given adequate health and safety information on the use of equipment, the dangers, etc. They must also be given suitable training.

Persons must be nominated to carry out maintenance and repairs on equipment, and must be suitably trained and competent.

Work equipment must be capable of isolation from any power source.

Maintenance operations are to be carried out when equipment is stopped, unless work can be done without exposure to health and safety risks.

Warning notices must be displayed as necessary adjacent to equipment.

Work equipment includes installations.

Mobile work equipment is included, as are woodworking machines and power presses.

Second-hand equipment brought into a business after 1 December 1998 will be classed as new equipment.

What are the key things I need to know about the Control of Substances Hazardous to Health Regulations 2002 (COSHH)?

Employers must carry out assessments of all work activities involving the use of substances hazardous to health which might pose health risks to employees or others.

All assessments must be suitable and sufficient and must be in writing.

Exposure of employees to hazardous substances must be prevented or otherwise controlled — the hierarchy of Risk Control must be followed.

Personal protective equipment is only permitted when other control measures are not practicable.

Control measures must be maintained in efficient working order and good repair. Ventilation equipment must be checked annually.

Monitoring of exposure of employees to substances must be carried out where appropriate.

Employees must be given health surveillance where appropriate and records kept for 40 years.

All employees exposed to hazardous substances must receive adequate information, instruction and training on the health risks created by their exposure to substances.

Information must be made available from manufacturers. Hazardous substances with OEL or MEL levels must be used in strict compliance with Codes of Practice, etc.

Safety data sheets must be attached to COSHH assessments.

What are the key things I need to know about the Fire Precautions (Workplace) Regulations 1997 (amended 1999)?

These Regulations require all employers to assess fire hazards within their workplace, i.e. produce Risk Assessments.

They apply to all employers irrespective of whether their premises have a Fire Certificate.

Appropriate controls must be implemented to either eliminate or control the risks.

Suitable provision must be made for fire fighting equipment.

All fire safety equipment must be maintained and tested regularly.

Emergency plans must be in place for raising the alarm in the event of a fire.

Staff must receive adequate training in fire safety matters.

There must be a fire detection system available.

Suitable means of escape in case of fire must be provided.

Fire Risk Assessment, procedures, etc. must be regularly monitored and updated when circumstances require.

What are the key things I need to know about the Pressure Systems Safety Regulations 2000?

These Regulations replace the Pressure Systems and Transportable Gas Containers Regulations 1989.

Transportable gas containers are now covered in the Carriage of Dangerous Goods (Classification, Packaging and Labelling) Regulations 1996 and the Use of Transportable Pressure Receptacles Regulations 1996.

The Pressure Systems Safety Regulations 2000 aim to prevent serious injury from the hazard of stored energy as a result of the failure of a pressure system or one of its components parts.

The Regulations are concerned with steam at any pressure, gases which exert a pressure in excess of 0.5 bar above atmospheric pressure and fluids which may be mixtures of liquids, gases and vapours where the gas or vapour phase may exert a pressure in excess of 0.5 bar above atmospheric pressure.

Pressure systems must be properly designed, must allow for proper inspections to be carried out and must have safe means of access.

Written information must be available regarding the design, installation, operation, examination, etc. of a pressure system.

Pressure systems shall be installed so that they are safe.

Pressure systems must have safe operating limits and the user must know what these are.

All systems must have a written scheme for examination and all systems must be examined in accordance with the written scheme.

If any pressure system demonstrates conditions of 'imminent danger' during an examination, the competent persons shall provide a written report specifying repairs, modifications or changes and shall within 14 days send a copy of the report to the enforcing authority.

All users of all pressure systems shall receive adequate and suitable instructions for:

- safe operating of the system
- actions to be taken in the event of an emergency.

Pressure systems must *not* be operated except in accordance with the instructions given.

All pressure systems shall be properly maintained in good repair so as to prevent danger.

No modifications or repairs shall give rise to danger.

Written records of examinations shall be kept, as well as any other relevant information.

What are the key things I need to know about the Confined Spaces Regulations 1997?

The definition of confined space covers, amongst others, the following:

- trenches
- vats

- silos
- pits
- chambers
- sewers
- vaults
- wells
- internal rooms.

Employers have a duty to ensure that their employees comply with the Regulations, and that they, as employers ensure that employees are not exposed to risks to their health, safety and welfare.

Any environment which could give rise to 'specified risks' can be covered by the Regulations. Specified risk includes:

- injury from fire or explosion
- loss of consciousness through a rise in body temperature or asphyxiation
- drowning
- injury from free flowing solids causing asphyxiation
- anything which prevents an escape from a space.

Risk Assessments are required for work in confined spaces.

Wherever practicable, work in any confined space shall be avoided.

Entry into a confined space is prohibited unless suitable rescue arrangements have been put in place.

Workers and rescuers must be trained in the hazards and risks associated with confined spaces.

What are the key things I need to know about the Control of Lead at Work Regulations 2002?

The Regulations apply to any type of work activity which is liable to expose employees and any other person to lead.

Lead is any lead, including lead alkyls, lead alloys, any compounds of lead and lead as a constituent of any substance, which is liable to be inhaled, ingested or otherwise absorbed by persons. Lead given off in exhaust fumes from road traffic vehicles is excluded.

Employers must:

- assess the risk to health from lead
- carry out Risk Assessments
- identify and implement measures to prevent or adequately control exposure to lead
- record the significant findings of the assessment
- protect employees where exposure is deemed to be significant
- issue employees with protective clothing
- monitor lead in air concentrations
- place employees under medical surveillance
- ensure high standards of personal hygiene
- provide employees with information, instruction and training
- identify the contents of containers and pipes
- prepare procedures for emergencies, accidents and incidents.

What are the key things I need to know about the Control of Asbestos at Work Regulations 2002?

Employers are responsible for the health and safety of persons other than their employees, whether at work or not.

Employers are not responsible for providing information, instruction and training to persons who are *not* their employees unless those persons are on the premises where the work is to be carried out.

A duty holder must manage any risks from asbestos and must arrange for assessments to be carried out to determine whether asbestos is, or is likely to be present, on the premises.

A duty holder is anyone who has a responsibility or contract for maintenance or repair of the premises, or who has control of any part of the premises.

More than one duty holder can exist for a premises and the level of responsibility will be determined by how much they have to contribute to repairs and maintenance.

Premises have to be inspected to determine whether asbestos is present.

Risk assessments must be made.

A premises register is developed and maintained as to the whereabouts of asbestos, how it is managed, etc. The register must be kept up to date and available to anyone who may need to see it.

No employer shall require any work on asbestos until its type has been identified.

No employer shall require any work on asbestos unless a risk assessment has been carried out.

Findings shall be recorded.

Asbestos types must be identified.

Control measures need to be assessed.

A suitable written plan must be produced for any work with asbestos.

The plan shall be kept on the premises for as long as the work continues.

Notification must be made at least 14 days before works are proposed to commence.

Every employer has to ensure that adequate information, instruction and training is provided for employees.

Information on the risk assessment must be given, including risks to health, precautions to be taken, control and action levels, etc.

Information, instruction and training must be given at regular intervals.

Employers must ensure that exposure to asbestos is prevented or, where this is not possible, reduced to the lowest practical level.

The number of employees so exposed must be as low as possible.

Substances may be used as a substitution for asbestos if they pose lesser risks.

Any control measures must be used correctly.

All control measures must be properly maintained, repaired, etc.

Ventilation equipment must be thoroughly tested and examined.

Suitable records of tests, maintenance, etc. shall be kept for at least five years.

Adequate and suitable protective clothing must be provided for employees working with asbestos.

Clothing so exposed to asbestos must be disposed of or cleaned.

Clothing must be removed from site, etc. in suitable containers.

If any personal clothing is exposed to asbestos it shall be treated as protective clothing and either disposed of or cleaned.

Emergency procedures shall be developed and tested, e.g. safety drills.

Information on emergency arrangements shall be available.

Suitable warning and communication systems shall be available.

Any unplanned release of asbestos shall be dealt with immediately so as to mitigate the effects, restore the situation to normal and inform persons who may be affected.

Every employer shall prevent or reduce to the lowest level reasonably practicable, the spread of asbestos from any place where work is carried out under his control.

Premises and plant shall be kept clean.

Premises, or parts, shall be cleaned after works are completed.

All employers shall ensure that any area where asbestos work is being undertaken is designated an 'asbestos area', or a 'respirator zone' where the level of asbestos exceeds the control level.

Notices designating these areas must be displayed.

All employees, other than those needing to work in the area, shall be excluded from the area.

No eating, drinking or smoking shall take place in a designated area, and other places for these activities must be available.

Asbestos fibres in the air shall be monitored by employers at regular intervals and when changes occur that may affect exposure.

Air monitoring is not required if exposure will be below the action level.

Records of air monitoring shall be kept.

Records shall be kept for 5 years, or 40 years if they form part of a health record.

Records should be available to those who wish to see them.

Employers must ensure health records are kept of employees exposed to asbestos unless exposure is less than the action level.

Records shall be kept for 40 years.

Medical surveillance shall be given every two years to employees exposed to asbestos.

Adequate washing and changing facilities must be provided to those employees exposed to asbestos.

All asbestos must be stored or transported in sealed containers properly labelled.

No person shall supply products containing asbestos.

An employer may exercise the defence of 'due diligence' or that he took all reasonable precautions to avoid the commission of the offence.

As an employer, am I responsible for the health and safety of non-employees?

Under Health and Safety Legislation, employers are responsible for the health and safety of persons who are not their employees or not in their employ.

The Courts consider the employment status of self-employed and casual/contract labour differently to that of the Inland Revenue. A person can be self-employed for tax reasons, but employed for health and safety reasons, especially where the employer instructs such persons as part of their business.

Three key legal cases confirm employers' responsibilities for others:

- *R* v. *Associated Octel* (Criminal Court)
- *Lane* v. *Shire Roofing Company* (Oxford) Ltd (Civil Court)
- *Nelhams* v. *Sandells Maintenance Limited and Another* (Civil Court).

Associated Octel

Associated Octel operated a chemical plant and instructed a specialist contractor to maintain and repair the plant during its annual shut-down.

During a repair operation, a light bulb operated by one of the Contractor's employees burst, lighting the cleaning fluid he was using and badly burning him.

Both the Contractor and Octel (the Client) were prosecuted, Octel for failing to ensure the safety of persons other than its employees.

Found guilty, Octel appealed to the Court of Appeal.

The Court of Appeal agreed with the initial Crown Court decision. They agreed that an employer's 'undertaking' can include the activities of a third party over which the employer had no control. The employer must show that it has done enough to protect the third party's safety.

The Court decided that an employer, by instructing a contractor to perform, would be 'conducting his undertaking' for the purposes of Section 3 of the Health and Safety at Work Etc. Act 1974, and as such has duties to ensure the safety of that contractor's employees.

Shire Roofing Company (Oxford) Ltd

Lane was a self-employed (for tax purposes) roofing contractor who was hired by Shire Roofing to re-roof a porch at a private house.

Lane told Shire that he could do the job for £200 'all in'. To make the job profitable he needed to use a ladder and not a scaffold or tower scaffold which was offered by the roofing company.

Shire Roofing did not get involved with Lane or the job, and did not supervise him, etc.

During the course of his work, Lane slipped and fell and suffered severed head injuries.

Lane brought a civil action for damages against Shire, claiming they had breached a common-law duty of care as an employer to ensure his health and safety as an employee.

The Court found that Mr Lane was an independent contractor and not entitled to any common-law duty by another.

He appealed to the Court of Appeal (Civil Division).

The Court of Appeal reversed the decision and stated that Lane had been hired as a labourer rather than as an independent contractor and that his tax status was irrelevant for health and safety purposes.

The Court of Appeal reviewed the 'control' test which the earlier Court had applied.

The control test obliges the Court to ask questions about the relationship between the parties, such as who decides what activities are to be performed and how they are to be carried out, who allocates the resources for the job and who sets the timetable, etc.

The Court of Appeal decided that the first 'control test' was inconclusive as it had failed to consider the professional skills and discretion that Lane had in making decisions about the work. When determining 'employment' the correct question was not 'who was in control', but rather 'whose business was it'.

Shire Roofing's supervisor felt that it was his responsibility to provide safety aids, materials and plant, etc. to short-term labourers, such as Lane.

The Court of Appeal decided that Shire Roofing had a duty of care to Lane and awarded him £102 000 damages, plus costs.

Sandells Maintenance Ltd

Nelhams was a permanent employee of Sandells.

Sandells 'loaned' Nelhams to another company — Gillespie, telling him he was under the 'complete control' of Gillespie.

Gillespie asked Nelhams to do some painting. He was told he would have to do the job from a ladder as scaffolding could not be used. Nelhams asked for the ladder to be footed but was told that no-one was available. While climbing the ladder, it slipped and Nelhams fell.

Civil claims were brought against Sandells and Gillespie by Nelhams. At trial, only Gillespie was found guilty as Sandells had handed over control of Nelhams.

Gillespie appealed to the Court of Appeal.

The Court of Appeal found that both employers were equally responsible for Nelhams' safety.

Even though there was an arrangement and 'loan' of an employee from one employer to another, the primary employer remains liable for the safety of his employees, despite the fact that he might loan them to another.

'The general employer cannot escape liability if the duty has been delegated and then not properly performed' — stated by the Court of Appeal.

Mr Nelhams was awarded damages, and both employers were found responsible.

But, the Court of Appeal then went on to rule that only Gillespie was wholly responsible for the accident as it failed to find labour to foot the ladder. Gillespie should therefore pay all the damages and should indemnify Sandells completely thereof.

What do these court cases mean?

An employer can now be prosecuted for failing to secure the safety of an independent contractor's employees, whether or not he has been involved with that contractor's activities.

An employer can be held liable for the safety of a person holding themselves out to be independent and self-employed if they instruct such persons as part of their business.

A temporary employer can be held financially liable for the injuries suffered by the employee of another.

3

Risk Assessments

What is hazard and risk in relation to health and safety?

Hazard means the potential to cause harm.
Risk means the likelihood that harm will occur.

A typical *hazard* on a construction site is falling from heights. The *risk* of falling is dependent on what control measures have been implemented, e.g.:

- the risk of falling is less if guard-rails are provided
- the risk of falling is less if safety harnesses are used
- the risk of falling is less if safe means of access is provided — stairs and not ladders, tower scaffolds and not ladders, etc.
- the risk of falling is less if materials do not need to be carried
- the risk of falling is less if wind and weather conditions are considered
- the risk of falling is less if adequate working space is provided, e.g. not perched on a ledge.

The *hazard* of falling from heights will be high but the *risk* of falling will be low if guard-rails are used.

Why are Method Statements/Risk Assessments to be completed and when should they be done and for what type of work?

Method Statements are written procedures which outline how a job is to be done so as to ensure the safety of everyone involved with the job, including persons who are in the vicinity.

A Method Statement equates to a 'safe system of work' which is required under the Health and Safety at Work Etc. Act 1974.

Risk Assessments are required under the Management of Health and Safety at Work Regulations 1999 and all employers are required to assess the risks to workers and any others who may be affected by their undertaking.

A Risk Assessment identifies hazards present, evaluates the risks involved and identifies control measures necessary to eliminate or minimise the risks of injury or ill-health.

A Method Statement can be used as the control measure needed to eliminate or minimise the risks involved in carrying out the job.

A Principal Contractor should carry out Risk Assessments for all work activities which *employees* undertake.

Also, a Principal Contractor, should carry out Risk Assessments for all those work activities which involve all operatives on site, i.e. communal activities — e.g. access routes to places of work, delivery of materials, plant and equipment, etc.

A Principal Contractor should receive Risk Assessments from all the other contractors, sub-contractors and self-employed tradesmen working on the site. These will detail what the hazards associated with their tasks are, e.g. noise from drilling equipment, and will include details of how the risks from the hazards, e.g. noise-induced hearing loss, can be eliminated or reduced.

When each of the contractor/sub-contractor Risk Assessments have been reviewed, the Principal Contractor must consider whether he needs to do anything else to protect other workers in the area, i.e. forbid certain work activities in certain areas, etc. If so, he will need to do an additional Risk Assessment which identifies how, as Principal Contractor, he is going to manage and control the combined risks of several contractors.

Risk Assessments need only identify *significant* risks involved in carrying out a work activity. Routine risks and everyday risks such as crossing the road to get to the employee car park need not be included.

Where anything unusual or uncommon is to be undertaken on the site, a Risk Assessment will be essential. Where works involve significant hazards, e.g. working in confined spaces, working at heights, working with harmful substances, then risk assessments are legally required and the control measures identified could be incorporated into a Method Statement which operatives are required to follow.

What are the key duties of an employer with respect to completing Risk Assessments?

The Management of Health and Safety at Work Regulations 1999 require employers to complete Risk Assessments for all activities undertaken at work for which there is a risk of injury or ill-health.

The Regulations place an absolute duty on employers to undertake suitable and sufficient assessments of:

- the risks to health and safety of employees to which they are exposed while at work

and

- the risks to health and safety of persons not in their employment arising out of or in connection with the conduct by them of their undertaking.

The key duties are:

- completing Risk Assessments
- ensuring that they are suitable and sufficient.

Any significant findings must be put in writing.

Employees must be given information, instruction and training on the Risk Assessments so that they understand the hazards and risks of the tasks to be undertaken. Employees must also know what control measures the employer has determined as being appropriate to minimise the risks of injury from the hazards identified.

What is a suitable and sufficient Risk Assessment?

A Risk Assessment should:

- identify the significant risks arising out of the work
- enable the employer to identify and prioritise the measures needed to be taken to ensure that all relevant statutory provisions are complied with
- be appropriate to the nature of the work
- remain in force for the duration of the work
- be regularly reviewed.

A Risk Assessment does not need to be perfect but must be 'the best which can be produced given the knowledge available at the time'. It must be 'proportionate' to the risks associated with the tasks.

Can anyone carry out a Risk Assessment?

There are no legally required qualifications for carrying out Risk Assessments. The Management of Health and Safety at Work Regulations 1999 require that whoever carries out a Risk Assessment must be 'competent' to do so.

Competency is not defined in any of the Regulations but guidance which accompanies various Regulations refers to 'appropriate knowledge, training and experience'.

The person carrying out the Risk Assessment must be able to:

- identify hazards
- judge the consequences of the hazards, i.e. likely injury
- determine how likely it is that the harm will be realised
- identify what control measures are needed to reduce the risks
- determine whether there is imminent risk of serious personal injury and prohibit the continuation of the task.

It would be sensible for the person carrying out the Risk Assessment to have had some training in the techniques and processes involved in conducting the Assessment.

Site Agents should be trained in the methodology of Risk Assessments.

Site Agents should enquire about the competency of any sub-contractors who purport to be experienced in carrying out Risk Assessments.

Under the Construction (Design and Management) Regulations 1994 any person who appoints a contractor must assess their competency and resources in respect of health and safety.

What are the five steps to Risk Assessment?

One of the best and most simple of guides to understanding Risk Assessment is the HSE's free leaflet:

Five Steps to Risk Assessment: INDG 163.

This very useful guide lists the five steps as:

Step 1: Look for hazards
Step 2: Decide who might be harmed and how
Step 3: Evaluate the risks and decide whether existing precautions are adequate or more should be done
Step 4: Record the findings
Step 5: Review and revise the Assessment if necessary.

Step 1: Hazard spotting

Walk around the site and look for anything which could cause harm. Ignore the trivial things and concentrate on those hazards which could cause serious harm.

Site hazards will include:

- trip hazards
- working at heights
- using electrical equipment
- manual handling
- noise
- harmful substances
- unguarded drops or voids
- easy access to site for unauthorised persons
- dust
- low temperatures
- vehicles and mobile plant
- insufficiently trained operatives
- using lifting equipment
- falling objects
- unstable structures
- tower scaffolds
- general scaffolding
- inadequate working platforms
- exposure to live services
- use of welding equipment
- likelihood of fire.

Step 2: Decide who might be harmed and how

Consider anyone who could be exposed to the hazard and who could be harmed as a result. Injury can be minor or major — but this really should not be a consideration as good health and safety is about preventing injury and ill-health, no matter how insignificant.

Consider, in particular, people unfamiliar with the site, namely:

- visitors
- delivery drivers
- clients
- short duration visiting contractors.

Also consider:

- young workers
- trainees and apprentices
- trespassers to the site
- members of the public
- anyone working on site, or visiting, with a disability
- women who may be pregnant (not an unusual event as more women are training in the building trades).

Step 3: Evaluate the risks and decide whether existing precautions are adequate

Having identified the hazards on site and who might be harmed, review what steps are currently in place to reduce the risk of injury. Usually, something is in place, either consciously or not, e.g. 110 v tools are used to reduce the risk of electric shock/electrocution from using 230 v tools.

Are the steps that are in place enough to prevent someone being injured at all?

These steps are called Control Measures.

If you think more could be done, then the Risk Assessment process requires you to implement additional controls where you can.

For example, you have identified a ladder as having the potential to cause harm — people could fall off. Their injuries could be severe. The current 'control' is to tie the ladder at the top to stop it moving. You have reduced the risk of people falling off the ladder due to the ladder moving unexpectedly.

But, you could do more to reduce the risk of falling from ladders altogether by requiring all operatives to use a mobile tower scaffold. This is a safer means of access than a ladder.

But, mobile tower scaffolds create their own hazards and so you will need to control these.

Have you done, are you doing, everything that is 'reasonably practicable' to keep the construction site safe?

Draw up an Action List of the controls which you could put in place. Consider the following:

- eliminating the hazard completely
- trying a less risky option
- preventing access to the hazard, e.g. guarding
- removing people from the hazard
- issuing PPE
- providing methods with which to deal with the hazard immediately, e.g. washing facilities to reduce skin contamination from substances.

Step 4: Record findings

It is always sensible to keep a record of what steps have been taken to address hazards and risks. If employers have five employees or more, Risk Assessments must be in writing, although you have only to record 'significant' risks.

This means writing down the significant hazards and your conclusions of who might be harmed, how and how often and what control measures are in place, or are to be put in place to control the risks.

These written records are often referred to as Risk Assessments. They do *not* need to be rocket science and they do *not* need to be perfect. Risk Assessments must be 'suitable and sufficient'.

You need to be able to show:

- that you reviewed hazards

- that you thought about what could happen, why, how and when
- that you considered who could be harmed and why
- that you considered all *known* steps that were practicable, which you could introduce or follow to reduce the hazards
- that the residual or remaining risk was lower than when you started
- that you informed people of the hazards and risks and what safety steps they should follow to reduce the hazards to acceptable levels
- that you recognise that the hazard and risk situation needs to be reviewed from time to time.

If one of the Statutory Enforcement Agencies visits site and notes hazards they will want to see, almost without fail, your Risk Assessments.

Keep Risk Assessments until at least the end of the project. If anyone has had an accident of any significance, keep the appropriate Risk Assessment for at least six years — remember an individual can launch a civil claim in the courts many years after they had the accident.

It is acceptable to refer to any other health and safety records, procedures, manuals, etc. in your Risk Assessment.

Risk Assessments do not have to be in any specific format, but must cover the information required.

Step 5: Review and revise the Assessment

Site conditions change frequently and the controls which were in place last week may no longer be appropriate to new working areas, procedures, etc.

Review the Risk Assessment:

- Has anything changed on site?
- Has the weather affected anything?

Case Study

The Principal Contractor was responsible for ensuring the delivery of materials to site. The delivery area incorporated the rear access road which was shared by a neighbouring retail premises. There were hazards to both the site operatives and adjoining tenants from the delivery vehicles and the off-loading of materials. Hazards included moving vehicles, restricted access to the roadway for emergency vehicles, off-loading materials from the lorries, dust, noise, falling objects. The risks from the hazards included being knocked over, being hit by materials, noise-induced hearing loss, breathing in dust and exhaust fumes, etc.

The Principal Contractor formulated the risk assessment, identifying the above as the hazards and risks and determining the control measures needed to eliminate or minimise the risks. These included: having a banksman to guide in the delivery vehicles, setting specific delivery times, liaising with the adjoining tenants, providing lifting devices, requiring engines to be switched off during delivery, avoiding reversing vehicles wherever possible, etc.

The Principal Contractor then prepared a short Method Statement which was given to the Site Foreman to follow when deliveries occurred.

The preparation of this Risk Assessment and Method Statement was the Principal Contractor's responsibility because he had overall management control of these activities and could co-ordinate everyone else's deliveries to site.

- Have new operatives started work?
- Have new openings been made or new work areas created?
- Has new plant been brought to site?
- Has site layout changed?

Check that the precautions you have introduced to reduce risk are working effectively. If you are having a number of accidents occurring on the site, that could indicate that controls are not working. Review and investigate all accidents and then review Risk Assessments.

What are 'site-specific' Risk Assessments?

HSE Inspectors prefer to see site-specific Risk Assessments on every construction site.

These are Risk Assessments which relate to the actual site conditions and the actual type of construction project and are not general or 'generic' Risk Assessments which address all sorts of hazards not really relevant to the site in question.

The value of Risk Assessments is that they consider the hazards and risks of an activity which employees, or others, are undertaking and they refer to actual site or workplace conditions which actually *exist*.

Many construction contractors have manuals of 'generic' Risk Assessments and often these are referred to as the site Risk Assessments. They are not.

The 'generic' Risk Assessments can be used as the basis of the Risk Assessment but the competent person must still review the actual site conditions to see if the hazards identified on Risk Assessment are less or more likely to cause harm due to the local circumstances. For example, the hazards of moving vehicles on a construction site are quite common and can be generically assessed but on your site, the risks of injury may be increased because of 'blind corners'. The generic Risk Assessment may not address the hazard of 'blind corners'. The HSE Inspector will expect the 'site-specific' Risk Assessment to do so.

Does one Risk Assessment satisfy all the Regulations?

Risk Assessments for various work activities are required under the following Regulations:

- Management of Health and Safety at Work Regulations 1999
- Manual Handling Operations Regulations 1992
- Personal Protective Equipment at Work Regulations 1992
- Health and Safety (Display Screen Equipment) Regulations 1992
- Noise at Work Regulations 1989
- Control of Substances Hazardous to Health Regulations 2002
- Control of Lead at Work Regulations 2002
- Control of Asbestos at Work Regulations 2002
- Dangerous Substances and Explosive Atmospheres Regulations 2002.

The Management of Health and Safety at Work Regulations 1999 are the overriding, superior Regulations and their requirements are superimposed over all the other Regulations.

A thorough Risk Assessment process under the Management of Health and Safety at Work Regulations 1999 will probably satisfy the requirements for all other Regulations, but other Regulations may contain specific control measures which will need to be considered, e.g. exposure to hazardous substances under COSHH and whether health surveillance is needed.

When considering the hazards of working in confined spaces, the requirements of the Confined Spaces Regulations 1997 need to be considered even though the Regulations themselves do not require specific Risk Assessments.

References

Management of Health and Safety at Work Regulations 1999.

Approved Code of Practice: Management of Health and Safety at Work Regulations: L21.

Five Steps to Risk Assessment: INDG 163.

A Guide to Risk Assessment Requirements: INDG 218.

4

Management of health and safety and the application of the CDM Regulations

What aspects of the Management of Health and Safety at Work Regulations 1999 apply to construction sites?

The Management of Health and Safety at Work Regulations 1999 (MHSWR) set out general duties for employers and employees in all non-domestic work activities and aim to improve health and safety management by developing the general principles set out in the Health and Safety at Work Etc. Act 1974.

MHSWR duties overlap with duties contained in several other pieces of health and safety legislation, including the CDM Regulations. Compliance with other legislation normally implies compliance with MHSWR but sometimes the duties in MHSWR go beyond those of other Regulations. In these instances, the duties imposed by MHSWR take precedence over others.

MHSWR places duties on employers (and the self-employed) including Clients, Designers, Planning Supervisors, Principal Contractors and other Contractors.

Under MHSWR, employers must:

- assess the risks to the health and safety of their employees and others who may be affected by the work activity (Regulation 3)
- identify what actions are necessary to eliminate or reduce the risks to health and safety of their employees and others
- apply the principles of prevention and protection
- carry out and record in writing, if they have five employees or more, a Risk Assessment
- make appropriate arrangements for managing health and safety, including planning, organisation, control, monitoring and review of preventative and protective measures. Arrangements must be recorded if employing five or more employees
- provide appropriate health surveillance for employees whenever the Risk Assessment shows it necessary, e.g. to check for skin dermatitis
- appoint competent persons to assist with the measures needed to comply with health and safety laws. Competent persons should ideally be from within the employer's own organisation. Where more than one competent person is appointed, the employer must ensure that adequate co-operation exists between them
- set up procedures to deal with emergencies and liaise, if necessary, with medical and rescue services
- provide employees with relevant information on health and safety in an understandable form
- co-operate with other employees sharing a common workplace and co-ordinate preventative and protective measures for the benefit of all employees and others
- make sure that employees are not given tasks beyond their capabilities and competence
- ensure that employees are given suitable training
- ensure that any temporary workers are provided with relevant health and safety information in order to carry out their work safely.

Employees have duties under MHSWR to:

- use equipment in accordance with training and instruction
- report dangerous situations
- report any shortcomings in health and safety arrangements
- take reasonable care of their own and others' health and safety.

The Principal Contractor will carry the bulk of the responsibility for MHSWR on a construction site and, as the site will be 'multi-occupied', the Principal Contractor must ensure co-operation and co-ordination between employers. This will be laid out in the Construction Phase Health and Safety Plan. Contractors must carry out their own Risk Assessments but the Principal Contractor must complete these where hazards and risks affect the whole workforce, e.g. site access routes, communal lifting operations. On a multi-occupied site, the Principal Contractor will assume overall responsibility for the management of health and safety and will co-ordinate and arrange emergency procedures, etc. Information on such procedures must be given to all persons using the site by the Principal Contractor. Information must be comprehensible and understandable so may need to be in picture form, cartoons, posters and foreign language, etc.

What are the five steps to successful health and safety management?

The five steps to successful health and safety management are exactly the same steps required for the successful management of any project, namely:

1. policy
2. organise
3. planning and implementing

4.   measuring performance
5.   reviewing performance.

Step 1: Set your own policy

All employers who have employees must have a Health and Safety policy and those who have five or more employees must have it written down and available for employees to consult.

All main contractors will have a Health and Safety Policy and this sets the framework for the management of health and safety on any site on which their employees are to work.

The Site Agent, Manager or Contracts Manager should review the Company Policy and see whether amendments or additions are necessary for the actual site in question. Does the main Policy reflect how you will manage health and safety on site? Have you identified your own safety objectives?

Step 2: Organise

Having identified the overall Health and Safety Policy and objectives for the site, the site and its operatives need to be organised to deliver the Policy and objectives. You will need to create a positive health and safety culture for the site — setting standards, enforcing standards, taking a strong lead, etc.

There are four components to a health and safety culture:

1.   competence — recruitment, training and advisory support
2.   control — allocating responsibilities, securing commitment, instruction and supervision
3.   co-operation — between individuals, other contractors and employers
4.   communication — spoken, written, visible.

In organising the site Health and Safety Policy and procedures have you:

- allocated responsibility for health and safety to specific people? Are they clear on what they have to do and are they held accountable?
- consulted and involved all operatives, contractors, sub-contractors, the self-employed and other employers, trade union representatives, etc.?
- given everyone sufficient information on the health and safety standards, objectives, hazards and risks of the site?
- ensured the right level of expertise on site to manage all tasks safely and effectively?
- a properly trained workforce and fully inducted operatives?
- specialist advice available to assist you in managing health and safety?

All of the above should be considered for each construction site before the work actually commences.

Step 3: Plan and set standards

If the CDM Regulations 1994 apply to the construction project, there must be a Construction Phase Health and Safety Plan before works start on site. Even if CDM does not apply to the construction project because it is too small, it is still a good idea to formulate a Construction Phase Health and Safety Plan.

What objectives do you want for the project?

Has the Client set these? Would any of the following be achievable:

- no major injury accidents
- lost working days to be no more than 1% of construction days
- Risk Assessments to be submitted 24 hours before works commence
- all operatives to be inducted to the Site Safety Standard within the first hour of commencement on day one
- no visitors to be let into the construction area unaccompanied

- all persons to wear, as a minimum:
  - ○ safety helmet
  - ○ safety shoes
  - ○ hi-vi vest.

Set some achievable objectives. Decide how you are going to monitor that you have achieved them. For example, how will you require all accidents to be recorded — or will you only require reportable accidents to be notified to you.

What mechanism will you introduce so that health and safety can be considered before any major work activity or site alteration.

What plans have you put in place for emergency procedures?

Step 4: Measure performance

Consider how you will monitor the performance of health and safety controls and standards.

Who will carry out safety checks and how often?

Will there be a sub-contractors' safety meeting?

Who will review contractors' Risk Assessments and Method Statements?

How will accident statistics be reviewed?

Who will worry about near misses?

Will the number of HSE inspections be recorded? Is there a system?

What is the lost time rate caused by accidents and ill-health?

Can any costs be quantified for lost time due to work stopping to investigate an accident?

What are the standards required by the Client and how well do you meet them, e.g. Permit to Work procedures?

Step 5: Audit and review

How well does the Health and Safety Plan work? Where can it be improved?

How do you learn from mistakes and successes?

If on-site safety checks are carried out, who implements action plans?

Can you review performance on this project against other projects?

Can you benchmark your health and safety compliance with competitors.

Health and safety management is a journey of continuous improvement, never be satisfied with 'making do' — the better the health and safety management, the safer the site and the more efficient and profitable the job. And the greater the contractor's reputation.

The CDM Regulations refer to key appointments which must be made on a construction project. What does this mean?

The CDM Regulations identify four key 'posts' which have responsibilities for ensuring that health and safety matters are addressed during construction projects. They are:

- The Client (or the Client's agent)
- The Designer
- The Planning Supervisor
- The Principal Contractor.

The Client is anyone for whom a construction project is carried out.

The Designer is anyone who carries on a trade, business or undertaking in connection with which he:

- prepares a design, or
- arranges for any person under his control to prepare a design,

relating to a structure or part of a structure.

The Planning Supervisor is a 'function' whereby the overall responsibility for co-ordinating health and safety aspects of the design and planning stage is undertaken.

The Principal Contractor must be a Contractor and must take responsibility for all site-specific safety issues, including ensuring that Contractors and Sub-contractors are competent and have resources to carry out the work safely and that a Health and Safety Plan is developed. Principal Contractors are also responsible for providing information, training and consultation with employees, including the self-employed.

When does CDM apply to a construction project?

The CDM Regulations apply to all construction work which falls into any of the following categories:

- lasts more than thirty (30) days
- involves, or is *expected* to involve, more than five hundred (500) person days
- involves *more than* four (4) people at work at any one time carrying out construction work
- involves *any* demolition or dismantling works.

The Regulations apply to:

- new build construction
- alteration, maintenance and renovation of a structure
- site clearance
- demolition and dismantling of a structure
- temporary works.

What is 'construction work' under the CDM Regulations and other Regulations?

Construction work means the carrying out of building, civil engineering or engineering construction work.
The definition includes:

- the construction, alteration, conversion, fitting out, commissioning, renovation, repair, upkeep, redecoration or other maintenance (including cleaning which involves the use of water or an abrasive at high pressure or the use of substances classified as corrosive or toxic) de-commissioning, demolition or dismantling of a structure
- the preparation for an intended structure, including site clearance, exploration, investigation (but not site survey), excavation, laying and installing the foundations of the structure
- the assembly of pre-fabricated elements to form a structure or the disassembly of pre-fabricated elements which previously formed a structure
- the removal of a structure or part of a structure or any product or waste resulting from demolition or dismantling of a structure or from the disassembly of pre-fabricated elements which, immediately before disassembly, formed a structure
- the installation, commissioning, maintenance, repair or removal of mechanical, electrical, gas, compressed air, hydraulic, telecommunications, computer or similar services which are normally fixed within or to a structure and which involve a risk of falling more than 2.0 m.

What is not construction work?

The following activities are generally not classed as construction work:

- general horticultural work and tree planting
- archaeological investigations
- erection and dismantling of marquees
- erection and dismantling of lightweight partitions to divide open plan offices
- creation of exhibition stands and displays
- erection of scaffolds for support or access for work activities which are not classed as construction works
- site survey works, e.g. taking levels, assessing soil types, examining structures
- work to or on ships
- on-shore fabrication of elements for off-shore installations
- factory manufacture of items for use on construction sites.

Are the CDM Regulations the only legislation applicable to construction projects?

No. A wide range of health and safety legislation applies to construction projects, especially where there are employers, employees and self-employed persons involved. Also, the laws of health and safety apply to 'persons in control of premises' and this could mean that the Client has some responsibilities for safety.

The Health and Safety at Work Etc. Act 1974 is the key piece of legislation applicable to all people 'at work' and to others who may 'resort' to premises.

Under the umbrella of the Health and Safety at Work Etc. Act 1974 are many subsidiary Regulations which could apply to construction projects. Some of these are:

- Noise at Work Regulations 1989
- Control of Asbestos at Work Regulations 2002
- Electricity at Work Regulations 1989
- Manual Handling Operations Regulations 1992
- Personal Protective Equipment at Work Regulations 1992

- Reporting of Injuries, Diseases and Dangerous Occurrences Regulations 1995
- The Construction (Health, Safety and Welfare) Regulations 1996
- The Health and Safety (Safety Signs and Signals) Regulations 1996
- The Fire Precautions (Workplace) Regulations 1997 and 1999
- The Confined Spaces Regulations 1997
- The Provision and Use of Work Equipment Regulations 1998
- The Lifting and Lifting Operations Regulations 1998
- The Control of Substances Hazardous to Health Regulations 2002
- The Management of Health and Safety at Work Regulations 1999
- The Construction (Head Protection) Regulations 1989
- The Control of Lead at Work Regulations 2002
- The Dangerous Substances and Explosive Atmospheres Regulations 2002

What health and safety legislation (other than CDM) is the most important on a construction project?

All health and safety legislation is important and failure to comply with any relevant legislation is an offence punishable by a fine and/or imprisonment. All members of the project team should be aware of which legislation applies to the project and, in particular, the Principal Contractor must have detailed knowledge of what is relevant to the site and proposed works.

The main piece of legislation applicable to construction sites is the Construction (Health, Safety and Welfare) Regulations 1996.

These Regulations set out general and specific requirements for maintaining health and safety on construction sites and cover issues such as:

- provision of welfare facilities
- working at height
- traffic routes and pedestrians
- working in excavations
- working on platforms
- safe places of work
- falling objects
- demolition and dismantling
- prevention of drowning
- stability of structures
- fire safety
- emergency procedures
- housekeeping
- training
- statutory inspections.

The Principal Contractor is responsible for ensuring health and safety on the construction site and must ensure that Contractors and Sub-contractors follow the requirements of the law. Where work activity on a site affects more than one Contractor, the Principal Contractor must take an overview and assess the risks to health and safety for all operatives.

Can anyone be appointed Principal Contractor?

The Principal Contractor is a legal appointment which the Client has to make under the CDM Regulations.

The Principal Contractor must be a Contractor, i.e. someone who either undertakes or manages construction work. A Client who normally co-ordinates construction works carried out on their premises, and who is competent and adequately resourced, can be a Principal Contractor.

The Principal Contractor would normally be a person carrying out or managing the construction work on the project to which they are appointed, i.e. the main or managing Contractor. However, where specialist work is involved, it may be appropriate to appoint the Specialist Contractor as Principal Contractor as they would be more suited to managing the risks of the specialist activity.

What are the duties of the Principal Contractor?

The Principal Contractor has specific duties under the CDM Regulations and, ultimately, carries responsibility for site safety issues. The main duties can be summed up as follows:

- develop and implement the Construction Phase Health and Safety Plan
- appoint only competent and properly resourced Contractors to the project, e.g. Specialist Contractors, Sub-contractors
- obtain and check Method Statements from Contractors
- ensure the co-operation and co-ordination of Contractors while they are on site, i.e. control multi-occupied site working
- ensure health and safety training is carried out
- develop appropriate communication arrangements between Contractors in respect of site health and safety issues
- make arrangements for discussing health and safety issues relative to the project
- allow only authorised persons onto the site
- display F10 on site for all operatives to be able to see the details
- monitor health and safety performance on site
- pass information to the Planning Supervisor for the Health and Safety File.

The Principal Contractor is a critical appointment to the construction project as the standard of site safety will be determined by the commitment and competency of the appointment.

When is the Principal Contractor appointed and how is this done?

The Client has to appoint the Principal Contractor as soon as is practicable after having information about the construction aspects of the project, and in good time so that the Client can determine the competency and resources of the Principal Contractor.

If the preferred Contractor is known prior to the tendering or negotiating phase of the project, he should be appointed early in the design process.

Usually, the Principal Contractor is appointed as a result of the tendering process and is usually the successful tenderer.

The appointment of Principal Contractor is often assumed, or is a verbal instruction but it is best defined in writing from either the Client, Contract Administrator or Planning Supervisor (if requested to do so).

The preliminaries in the Bill of Quantities may stipulate that the successful tenderer will be appointed Principal Contractor. Equally, so can a statement be made in the Pre-tender Health and Safety Plan.

The timing of the appointment should allow the Principal Contractor opportunity to develop the Construction Phase Safety Plan.

What are the key duties of the Principal Contractor prior to the commencement of works on site?

The main duty of the Principal Contractor is to develop the Pre-tender Health and Safety Plan into a Construction Phase Health and Safety Plan. This must be done prior to the commencement of works on site.

The Principal Contractor, if appointed early enough, can add great value to the design process and should be consulted about construction sequences, methodology, available and suitable materials, etc.

If the Principal Contractor has any design responsibilities, he must provide Design Risk Assessments to the Planning Supervisor where appropriate and must ensure that he co-operates with all other designers.

The Principal Contractor should have visited the site prior to the commencement of works and should liaise with the Planning Supervisor regarding any site-specific hazards and risks identified in the Pre-tender Health and Safety Plan.

Prior to the commencement of construction, the Principal Contractor should have identified and, if necessary, have appointed, key Contractors whom he believes to be competent and with the right resources to do the job. Regulations 8 and 9 of CDM state that 'no person shall arrange for a contractor to carry out or manage construction work unless he is reasonably satisfied that' the contractor has the competence and resources to undertake the work and comply with legislation.

The requirement to assess competency and resources does not just rest with the Client — everyone who appoints Contractors has to satisfy themselves of their competency and resources.

If the Principal Contractor appoints the mechanical and electrical contractors then they must ensure their competency and this is best done prior to the commencement of construction works. Remember — 'every moment in planning saves three or four in execution'.

Can there be more than one Principal Contractor on a project to which CDM applies?

There can only be one Principal Contractor at any one time on a project, although there may be many main Contractors undertaking the works.

The Principal Contractor is a specific post required under the CDM Regulations with responsibility for managing and co-ordinating all of the Construction Phase Health and Safety Plan issues. The Principal Contractor must be given overall responsibility for co-ordinating construction phases and, as this is such an important function, there can only be *one* such appointment.

Notwithstanding the above, at the same location there could be one or more different projects being carried out for different Clients and in these circumstances, one or more Principal Contractors could be appointed. Projects have to be distinct from each other and must not rely on one another for their viability and completion. If projects share common entrances or site access, need communal lifting equipment, share services to site, etc. then it is preferable to appoint one Principal Contractor with overall site management and co-ordinating responsibility so that these 'common resources' can be managed for the benefit of the site.

Each construction project could still have a main Contractor responsible to their Client but a representative should liaise with the Principal Contractor. Also, Planning Supervisors need to co-operate so that they can relay information to the respective Contractors regarding site and design hazards and risks generated by other parts of the project.

What is the Construction Phase Health and Safety Plan?

The Construction Phase Health and Safety Plan is the document produced by the Principal Contractor which develops the Pre-tender Health and Safety Plan prepared by the Planning Supervisor.

The two documents make up the project's Health and Safety Plan as required by Regulation 15 of CDM.

The Health and Safety Plan is the foundation upon which the health and safety management of the construction phase needs to be based. A written Plan clarifies who does what, who is responsible for what, what hazards and risks have been identified, how works shall be controlled, etc.

The Contractor appointed Principal Contractor must develop the Pre-tender Plan *before* construction works start so that it outlines the health and safety procedures which will be adopted during the construction phase.

The Construction Phase Plan must be site-specific, i.e. it must cover issues which apply to the works to be carried out, include actual site personnel, site-specific emergency procedures, etc.

Regulation 15(4) of CDM lists the information which must be included in the Construction Phase Plan as follows.

- Arrangements for the project (including, where necessary, for management of construction work and monitoring compliance with the relevant statutory provisions) which will ensure, so far as is reasonably practicable, the health and safety of all persons at work carrying out the construction work and all persons who may be affected by the work of such persons at work, taking account of:
  - ○ the risks involved in the construction work
  - ○ any activity of persons at work which is carried out, or will be carried out, on or in premises where construction work is undertaken
  - ○ any activity which may affect the health and safety of persons at work or other persons in the vicinity.
- Sufficient information about arrangements for the welfare of persons at work by virtue of the project to enable any Contractor to understand how he can comply with any requirements placed upon him in respect of welfare by or under the relevant statutory provision.

A Construction Phase Health and Safety Plan outline is shown in Appendix F.

The Health and Safety Plan must remain in existence and must be relevant until the end of the construction works. It may need to be updated as works progress and site conditions change.

What format does the Construction Phase Health and Safety Plan have to be in?

The Construction Phase Health and Safety Plan should be in a format which is:

- easy to use and to refer to
- understandable to those who need to use it
- easy to update
- easy to duplicate
- clear, concise and logical.

The Construction Phase Health and Safety Plan is needed *on site* and should therefore be robust and almost non-destructible. An A4 ring binder is a popular choice for keeping the information in order, as pages can be easily removed and photocopied for other Contractors as necessary and, importantly, it can be easily updated.

The Construction Phase Health and Safety Plan would not be particularly useful on computer disk on the site because access to its information might be restricted to the few.

The Construction Phase Health and Safety Plan need not be all written words — often the use of diagrams, pictograms and cartoons are very effective at explaining health and safety messages.

The Plan should not contain every health and safety procedure known to man, but only those *applicable* to the site. This should prevent it from becoming unwieldy and will not discourage people from reading it. It is perfectly acceptable to cross-reference to other documents within the Plan, e.g. the detailed Risk Assessment Manual, or to the COSHH Manual. If site safety procedures are reliant on safety procedures specified in other documents then copies of these must be available on site.

Does a copy of the Health and Safety Plan have to be given to every person working on the site, the Client, the Planning Supervisor, etc.?

No. The Client must be given a copy of the Plan so that they can be satisfied that it has been prepared and complies with the Regulations.

There is no duty to give a copy to the Planning Supervisor unless the Planning Supervisor is acting instead of the Client in assessing its adequacy before construction works can start. In this case, there would be no need to send one to the Client.

The CDM Regulations require the Principal Contractor to provide information to all persons working on or resorting to the site in respect of health and safety issues.

As the Health and Safety Plan contains valuable information on site health and safety matters, it makes sense to issue the document to as many people as practicable. However, that may become expensive and some operatives may only be on site for a few days.

Site safety rules should be issued to all individual operatives and could be issued to all employers/Contractors/employees during site induction training.

Specific Risk Assessments should be issued to the ganger or foreman, with instructions that he is responsible for ensuring that all his gang/team are made aware of hazards and risks and the protective measures needed to control the risks.

Copies of relevant information could be displayed in the mess room, site office and site canteen, e.g. location of first aid kit, names of companies with trained first aiders.

Key aspects of the Health and Safety Plan can be issued to safety representatives, site foremen, etc. with guidance on how and where they can access the full Health and Safety Plan and supporting information and documentation, e.g. Company Safety Policy, HSE Codes of Practice, etc.

A practical way of disseminating site health and safety information is to convene a weekly site safety forum or committee, requiring a foreman or representative from every Contractor or self-employed person on site to attend, using the meeting to review site health and safety issues and to discuss forthcoming works on the programme, new site safety rules, etc.

What is the best way to update the Construction Phase Health and Safety Plan without it becoming complicated or confusing?

The Construction Phase Health and Safety Plan is the document which sets out the health and safety management of the construction phase of the project. It must be a document which outlines what special health and safety precautions are to be taken to ensure the safety of everyone on the construction project and must be updated to reflect any changes to the working procedures, management systems, welfare facilities, etc., which happen on the site.

If the majority of the Construction Phase Health and Safety Plan has been agreed before the commencement of the construction works there will be little need to change substantial parts of it. Details on site management, emergency procedures and welfare facilities may not change during the construction phase if they have been well thought through at the beginning.

If the Construction Phase Health and Safety Plan needs to be updated it should be done by adding information clearly and removing old information so as to avoid confusion. For instance, the names of the trained first aiders may change. The new ones should be added to the Plan and the old ones removed. If the location of the first aid kit has changed this should be included.

The most important thing about the Construction Phase Health and Safety Plan is that the information contained in it is made available to all operatives on site — the simpler the updates, the it will be to understand.

Risk Assessments and Method Statements could be included at the back of the Plan, making it easy to add new information.

If the Principal Contractor decides to implement a new Permit to Work procedure for a specific activity which has only recently come to light, then this Permit to Work system must be clearly explained in the Construction Phase Health and Safety Plan.

An aspect of the Plan which will need to be kept under constant review, and updated when necessary, will be the Fire Safety Plan. As construction work progresses, site exit routes may be altered, e.g. by permanent partitioning, etc. Alterations must be clearly depicted on the Fire Safety Plan.

The Principal Contractor should ensure that, perhaps once a week, time is set aside to review the Construction Phase Health and Safety Plan and any relevant changes which are made must be *communicated* to site operatives via the arrangements made for that ensuring health and safety issues are considered.

Feedback from site operatives on health and safety matters should be considered and the Construction Phase Health and Safety Plan amended or updated to take into account operatives' concerns, ideas and suggestions as to how the site could be improved from a health and safety point of view.

Generic Risk Assessments will need to be reviewed and updated to incorporate site-specific issues. These should then be kept in the Appendix of the Construction Phase Health and Safety Plan, together with any associated Method Statements. Individual Risk Assessments can be issued to specific operatives as necessary or, importantly, to the Contractor Foreman so that he can assess what safety precautions need to be followed by his team.

What criticisms does the Health and Safety Executive have of Construction Phase Health and Safety Plans?

The HSE have raised many concerns about the quality of Construction Phase Health and Safety Plans, in particular regarding the general content which is often not relevant to the project in hand. They would prefer thinner but more site-specific documents.

Some of the common deficiencies are itemised as follows.

- Activities not assessed, i.e. those activities with health and safety risks which affect the whole site or specific trades (e.g. storage and distribution of materials, movement of vehicles, pedestrian access ways, removal of waste, provision and use of common mechanical plant, provision and use of temporary services, commissioning and testing procedures, etc.).

- Management arrangements do not focus sufficiently on the role of Risk Assessments.

- Site supervisors and managers do not have reasonable knowledge of safety, health and welfare requirements and standards.

- Site supervisors and managers are not familiar with the contents of the Construction Phase Health and Safety Plan.

- Monitoring arrangements are overlooked or the 'competent' person performing this role is not suitably qualified.

- Details of welfare provision is limited to a couple of lines of the Plan. It should cover in explicit detail the requirements and implementation of Schedule 6 to the Construction (Health, Safety and Welfare) Regulations 1996.

- Fire precautions, including arrangements for the fire alarm system (if required) and emergency lighting are often overlooked.

- The implication for health and safety of tight time-scales for the project are not fully addressed in the Plan. The Plan often fails to recognise that shortening a construction programme increases the amount of material stored on site and increases the number of operatives on site, all of which leads to restricted work space, inadequate supervision, poor co-ordination and control, etc.

The HSE Inspectors believe that all of the above must be considered before works commence on site and that if a Construction Phase Health and Safety Plan does not adequately address them, a Client should not approve the document under Regulation 10 of the CDM Regulations.

Can generic Risk Assessments be used as the basis of the Health and Safety Plan?

Yes, but they may not be sufficient for you to demonstrate that you have done everything 'reasonably practicable' to ensure the health and safety of all persons at work carrying out construction works.

Generic risk assessments, i.e. those which cover the general work activities (e.g. bricklaying), form the basis of identifying hazards and risks associated with the job. Provided that you develop the generic risk assessment to include any site-specific hazard (e.g. carrying out brickwork adjacent to a deep water course), and the additional control measures you intend to adopt, you will have sufficient information to ensure that operatives work safely.

When reviewing a Construction Phase Health and Safety Plan, the HSE Inspector will not be satisfied with a Plan which contains only general information, no matter how thick and impressive the Plan looks. Often, a much thinner and more accurate site-specific Plan will gain praise from the Inspector. They will look for information proportional to the project risks — too much information is confusing and off-putting.

What other requirements in respect of CDM does the Principal Contractor have to comply with?

In addition to developing the Construction Phase Health and Safety Plan, the Principal Contractor has specific duties laid down in Regulation 16 of the CDM Regulations.

These include:

- ensuring co-operation between all Contractors on the site or on adjacent sites where there is an overlap (e.g. shared access routes)
- ensuring that all Contractors and all employees work in connection with the rules contained in the Health and Safety Plan
- ensuring that only authorised persons are allowed into the premises where construction works are being carried out
- ensuring that the HSE notification of Form F10 is displayed, is in a readable condition and is in a position where anyone involved in the construction works can read it

- providing the Planning Supervisor with information, particularly if that information would be necessary for inclusion in the Health and Safety File
- giving reasonable direction to any Contractor, so far as is necessary, to enable the Principal Contractor to comply with his duties
- ensuring that the safety rules contained in the Health and Safety Plan are in writing and brought to the attention of persons who may be affected by them.

In order to be able to ensure co-operation between all Contractors it is necessary to have an understanding of Regulations 11 and 12 of the Management of Health and Safety at Work Regulations 1999. These Regulations require employers and the self-employed to co-ordinate their activities, co-operate with each other and to share information to help each comply with their statutory duties.

For instance, to be effective, Risk Assessments will need to cover the workplace as a whole and the duty to co-ordinate these will be the Principal Contractor's. Information must be provided by all employers/Contractors so as to enable the Principal Contractor to co-ordinate activities.

Another important aspect of co-ordination and co-operation relates to the use of work equipment and tools which are shared by all Contractors on the site. The Principal Contractor may assume responsibility for provision, maintenance and testing of all common equipment (e.g. lifting devices) or he may pass the responsibility on to another Contractor. It does not matter *who* does it as long as *someone* assumes responsibility and everyone else knows who that someone is.

The Principal Contractor will need to request the names of all the people, Contractors, Clients, and the Design Team, etc. who wish to visit the site as 'authorised persons'. The Principal Contractor may authorise them to enter all or part of the site. The Principal Contractor should adopt a formal signing-in procedure which all persons should follow. Unauthorised visitors should be accompanied around the site by a trained operative.

Under Regulation 17 of CDM, as Principal Contractor you have responsibility for ensuring that every Contractor is provided with comprehensive information on the risks to health and safety to all employees and others on the site.

In addition, you must ensure that every employer carries out suitable training for all employees, relative to the works involved and, also, that they provide information relating to health and safety issues.

'Comprehensive' information need not just be in writing — it could be diagrams, drawings, or it could be information in languages other than English.

The most appropriate way to ensure information is available is to include it in the Health and Safety Plan.

Under Regulation 18 of CDM, as Principal Contractor you must ensure that there are procedures in place for any employee or self-employed person to discuss health and safety issues and that there are arrangements for co-ordinating the views of others in respect of health and safety issues.

Depending on the size of the project, all that may be required will be an item for health and safety on a site meeting agenda and a formal process whereby someone can raise health and safety concerns without fear of reprisals.

What training must be provided for operatives on site?

The CDM Regulations do not actually require the Principal Contractor to provide training (other than to his own employees) to operatives on site but to ensure that every Contractor is provided with comprehensible information on the risks to health and safety from work activities on the site.

Comprehensive information is information which is understood by everyone. It is meaningless to issue complex site rules and Risk Assessments if the understanding of written English is poor. Verbal instructions, diagrams, etc. may be more comprehensible.

Site induction training is considered to be the responsibility of the Principal Contractor and the information given should include:

- site rules
- emergency procedures
- fire safety
- accident procedures
- Permit to Work systems
- site security
- welfare facilities on site
- first aid facilities
- management of health and safety on the site.

All the above information should be included in the Construction Phase Health and Safety Plan and a copy of the Plan should be given to the site foreman of every Contractor on the site.

Whether the Principal Contractor or each individual Contractor carries out site induction training, it is essential that written records of the training are kept and regularly reviewed and updated.

The Principal Contractor may require all Contractors to provide evidence of competency in the various trades and will be entitled to request training certificates for trades such as mobile equipment driving, fork-lift truck driving, gas-fitting works, etc.

All such documentation should be kept readily available by the Principal Contractor and is best kept appended to the Construction Phase Health and Safety Plan.

The Health and Safety Plan should set out what level of training site operatives are expected to have, who is to have provided it, how often and to what standard.

The Principal Contractor's role in respect of training is a co-ordination role, unless he is an employer of his own workforce when the requirements of health and safety training and the provision of information will apply equally to the Principal Contractor as to others.

The Client has appointed several 'Client Direct' appointments who are employed by the Client and required to access the site to undertake works. Are they exempt from complying with CDM?

No. As Principal Contractor you have absolute responsibility for site safety issues and can specify these in site rules included in the Construction Phase Health and Safety Plan.

Even if the 'Client Direct' is not a 'Contractor' under the Regulations, they will either be employers or employees and, as such, have legal duties under the Management of Health and Safety at Work Regulations 1999 to comply with the requirements of the Principal Contractor.

Contractors are defined in Regulation 2(1) of CDM as:

> Any person who carries on a trade, business or other undertaking (whether for profit or not) in connection with which he:
> * undertakes to or does carry out or manage construction work
> * arranges for any person at work under his control (including, where he is an employer, any employee of his) to carry out or manage construction work.

Construction work includes fitting out, commissioning, alteration, conversion, etc.

Regulation 19(2) of CDM applies to all *employers* and states that they must not cause or permit any employee to work on construction work unless the employee has been provided with relevant information, namely the name of the Planning Supervisor and Principal Contractor and the contents of the Health and Safety Plan. In any event, Regulations 11 and 12 of the Management of Health and Safety at Work Regulations 1999 apply, i.e. co-operation of a multi-occupied site and the appointment of a controlling employer for the site. Information has to be shared regarding Risk Assessments as required under Regulation 10 of the Management of Health and Safety at Work Regulations 1999.

The Principal Contractor has a duty under Regulation 16 of CDM to ensure that every *employee* at work in connection with the project complies with any rules contained in the Health and Safety Plan. If the Health and Safety Plan contains a rule which states that all Client Direct appointments, and any others conducting a trade, business or undertaking, shall provide Method Statements/Risk Assessments to the Principal Contractor prior to commencing work, they must legally comply with the rule.

What actions can be taken where either Contractors or Client Direct appointments fail to comply with the Site Rules, Health and Safety Plan or CDM requirements?

Remove them from site — probably more easily said than done.

Write to the employer of the Contractors/Client Direct appointments and advise them that they are in breach of their duties under CDM Regulations and other Health and Safety legislation and unless they start complying they will be removed from site, incurring any contractual penalties.

Often, non-compliance with a requirement is due to fear or ignorance. Perhaps they do not know how to conduct Risk Assessments. If this is the case, provide information and guidance.

Review your procedures for assessing competency of Contractors. Remember the duty to ensure competency of Contractors rests with *any person*, including Principal Contractors, who lets subsidiary work packages to Sub-contractors. Did you know the Contractor could not provide Risk Assessments? If so, why appoint them.

Consider whether you are asking the Contractor to provide more information than is justifiable. Risk Assessments and Method Statements need to be relevant and cover *significant* risks. Requesting meaningless paperwork from Contractors merely compounds their reluctance to produce any.

Ask the Planning Supervisor to help in encouraging the Contractor/Client Direct appointment to comply with their statutory duties. The Planning Supervisor could give advice to the Client that the Contractor is not competent and recommend that the Client moves to formally dismiss the Contractor.

Ask the local HSE Inspector for guidance. If the site is complying with CDM then there is nothing to fear in seeking advice from an Inspector on how to improve your health and safety management procedures (e.g. improving Contractor compliance).

De-list the Contractor/Client Direct appointment from your approved list.

Whatever steps you take, do not allow the breach of safety management procedures to go unrecorded. Keep detailed records of what actions you took to ensure compliance with site rules, who you spoke to, when, how often, etc. Make sure you have given them the information they could reasonably expect to have regarding the site works (e.g. Health and Safety Plan).

What notices under CDM must the Principal Contractor display on site?

The Principal Contractor must ensure that a copy of the Notification of Project (Form F10) as sent to the Health and Safety Executive by the Planning Supervisor is displayed on the site so that it remains legible and can be read by those working on site.

The Notice can be displayed in the Site Office but this may restrict the number of operatives who could easily refer to it. An acceptable place to display the F10 would be in the welfare or messing facilities where it was readily available to all operatives. A copy could be kept in the Site Office or with the Construction Phase Health and Safety Plan.

The location of the F10 Notice needs to be brought to the attention of all Contractors working on the site and the easiest way to do this is to inform them during the site induction training.

What subjects would be appropriate to include in induction training organised by the Principal Contractor?

The duty of the Principal Contractor is to convey information on risks relating to the carrying out of construction works within the designated site. One of the most effective ways to communicate information is via a structured training programme. Subjects to cover in induction training would be:

- outline of the project — who's who, Client, Design Team, etc.
- statement on health and safety and commitment of senior management to high standards of health and safety
- site-specific risks, e.g.:
  - access routes
  - contaminated land
  - overhead power cables
  - underground services
  - proximity of water
  - unstable buildings
  - hazardous materials (e.g. asbestos)
- site-specific control measures for identified risks
- site rules
- welfare facilities available, location, maintenance and cleaning provision, etc.
- first aid facilities
- accident and near miss reporting procedures
- emergency procedures, e.g.:
  - fire evacuation
  - raising the fire alarm
  - name and address of emergency services
  - assembly point
  - fire marshals/wardens
  - building collapse
  - flood, chemical escape, gas escapes
  - release of hazardous substances (e.g. asbestos)

Case Study

A new university campus was being constructed for a consortium of developers. The developers nominated the main developer as Client under the CDM Regulations 1994 and, as Client, the developer appointed a Management Contractor to oversee the entire construction works. The Management Contractor was designated Principal Contractor even though they were not undertaking any actual construction works but because they managed the construction process they met the qualification requirement of CDM for a Principal Contractor.

The Client required the Principal Contractor to have overall responsibility for the site and, although each tenant shop fit-out had a main contractor, only the Management Contractor was designated Principal Contractor.

The Client imposed a duty on the Principal Contractor to carry out induction training for *all* persons entering the site.

The Principal Contractor set up a separate training room within his site office compound and, via a strict security control point, all persons entering the site for the first time had to report to the training office for induction.

Each induction programme ran for 30 minutes and three sessions were undertaken each day — two in the morning and one early in the afternoon. If persons wanted to gain access to the site at other times they had to wait until the next induction session. Every person going through the induction process was registered and when they had completed the course they signed a declaration to that effect. These records were kept centrally by the Management Contractor.

Each attendee received a photo ID card which indicated that they had been inducted and when. On the reverse of the ID card were the basic emergency rules of the site.

Each individual main contractor undertaking their own Client's shop fit-out was required to provide additional training to operatives about the specific hazards and risks found on the site.

Each main contractor had to regularly provide information to the Principal Contractor on any health and safety issue on their site which could affect the safety of the whole site (e.g. LPG storage). The Principal Contractor then ensured that this information was added to its induction training programme.

- responsible persons
- requirements for protective equipment and clothing, e.g.:
  - use of hard hats
  - use of safety footwear
  - use of ear defenders
- general site safety controls, e.g.:
  - Permit to Work
  - Permit to Enter
  - use of banksmen
- arrangements for communicating with all the workforce in respect of health and safety, e.g.:
  - weekly site safety meetings
  - notice board
  - display of safety notices
  - use of other aids (e.g. visual aids to assist those with language difficulties)
- names of safety representatives, competent persons, etc.
- site security and access procedures.

All of the above information should be available in the Construction Phase Health and Safety Plan and key details should be displayed on an Information Board at the entrance to the site or in the messing facilities.

What are the main duties of contractors under the CDM Regulations and other construction legislation?

Contractors are given specific duties under Regulation 19 of CDM as follows:

- co-operate with the Principal Contractor in order that they both comply with their legal duties under *all* applicable legislation

- provide the Principal Contractor with information, including if necessary, Risk Assessments, on any activity, material, process or task which might affect the health and safety of any person at work carrying on construction works, or of any person who may be affected or which might cause the Principal Contractor to review health and safety across the site
- comply with any directions the Principal Contractor gives in relation to site rules or any other health and safety matter which the Principal Contractor has a duty to fulfil
- comply with the rules contained in the Health and Safety Plan
- provide information promptly to the Principal Contractor in relation to accidents, diseases or dangerous occurrences, including any fatalities, which the Contractor would need to report under the Reporting of Injuries, Diseases and Dangerous Occurrences Regulations 1985
- provide information to the Principal Contractor on any matter which it might be reasonable to assume the Principal Contractor will need to pass onto the Planning Supervisor for the Health and Safety File.

In addition to the above, if the contractor is an employer or a self-employed person they shall not start any construction works until they have been provided with the name of the Planning Supervisor and the Principal Contractor (i.e. have been given, or have seen, a copy of Form F10).

All contractors must have relevant parts of the Construction Phase Health and Safety Plan as relate to the work they will carry out on the site.

Contractors are employers (or self-employed) and therefore all health and safety legislation applies to them. It is *not* the Principal Contractor's responsibility to ensure that contractors comply with the law, but their own responsibility. The over-riding duty of all employers is to ensure that they protect the health, safety and welfare of their employees while they are at work.

Employers also have a duty to protect others affected by their undertaking and so this means the Principal Contractor and his employees, other contractors, self-employed people, visitors, Clients and members of the public.

No contractor on a construction site should do *anything* which will materially affect the health, safety and welfare of other people. They must identify the hazards and risks and implement controls to reduce the likelihood of the risks occurring.

What responsibilities do contractors have for on-site training of their operatives?

Where contractors are employers they have general duties under health and safety legislation to ensure that their employees receive suitable and sufficient training, or information, instruction and training.

A general duty for training employees is contained in Regulation 13 of the Management of Health and Safety at Work Regulations 1999, especially where they are likely to face new or increased risks within the work environment.

The Principal Contractor has a duty under CDM to ensure that contractors provide training to their employees and could include these requirements in the Construction Phase Health and Safety Plan.

Principal Contractors may require evidence from contractors of such training in, for example, the form of training records, copies of certificates, etc.

Contractors must ensure that they address any specific training needs relative to the site activities and, in particular, they must cover topics which their employees may not normally be familiar with.

Case Study

A contractor was appointed by the Principal Contractor to provide specialised finishes to internal walls and ceilings. The Principal Contractor had arranged for all works at height to be carried on from mobile elevating work platforms for safety reasons.

The specialist contractor's operatives were not familiar with using these devices as their work usually involved using mobile tower scaffolds.

The Principal Contractor gave the contractor's operatives induction training regarding the general hazards on the site, emergency procedures, etc. and explained the management process for communally used equipment, provided by the Principal Contractor.

Prior to allowing works to commence, the Contracts Manager of the specialist contractor, arranged for his operatives to receive additional training on the safe use of mobile elevating work platforms.

Do contractors have to provide Risk Assessments to the Principal Contractor?

The Principal Contractor has a duty to co-ordinate health and safety across the construction site and, in particular, must address hazards and risks which affect all operatives on the site, no matter who their employer.

Under Regulation 19, a contractor must provide relevant information to the Principal Contractor on any activity which might affect the safety of operatives or others working on or resorting to the site.

Relevant information includes any part of a Risk Assessment made under the general provisions of the Management of Health and Safety at Work Regulations 1999.

Principal Contractors should not really require Risk Assessments for tasks which only affect the contractor's operatives but often, in order to ensure that a culture of health and safety pervades the site, the Principal Contractor may include a Site Rule which states that Risk Assessments for all activities must be provided to them. Contractors have a duty to comply with site rules.

What are some of the practical safety initiatives which can be introduced to the construction project by the Principal Contractor?

Site safety does not stop with compliance with the CDM Regulations 1994. Legislation really only sets minimum standards to be achieved and good safety management endeavours to achieve higher standards. Good safety management has been proven to have positive benefits to employers, contractors, Clients and employees and ultimately reduces costs due to accidents, incidents, down time, stoppages, investigations, loss and damage to property, equipment, etc.

In order to encourage high standards of safety on construction sites, the Health and Safety Executive launched, during Summer 2000, a campaign known as Working Well Together.

This campaign encourages all those involved in a construction project to work together to reduce hazards and risks and improve overall standards of health and safety in construction.

The following sections outline some of the initiatives which have been or can be introduced during a construction project.

Preventing falls from heights

The Principal Contractor undertook the following actions:

- influenced designers to schedule permanent access structures, e.g. staircases, early in the project
- provided communal mobile tower scaffolds for use by all contractors and assumed responsibility for daily checking and remedial repairs by competent persons
- provided full edge protection to all drops of more than 2.0 m
- included within the site rules procedures for safe working at heights
- introduced a Permit to Work system for all work to be undertaken above 2.0 m
- required all contractors to provide Risk Assessments and a work plan for all activities they proposed to undertake above 2.0 m
- required all contractors to undertake tool-box talks on working at heights, the hazards, risks and controls necessary and required evidence from each contractor of such training or instruction
- provided a site audit function to check regularly on activity.

Preventing slips, trips and falls

The Principal Contractor introduced the following:

- information on slips, trips and falls, types of injury sustained, etc. included with site induction training

- clearer emphasis within site safety rules on good housekeeping, trailing cables, etc.
- provision of adequate number of transformers, etc. in locations near to where power tools are to be used
- regular site safety inspections
- employed a site labourer dedicated to clearing the site of debris, etc.
- adequate provision for skips, etc. on site for containing waste materials
- regular maintenance of floor surfaces, infilling of holes, etc.
- specific tool-box talks with electricians so that they did not leave floor box recesses unguarded
- provision of good, overall site lighting and, where necessary, appropriate task lighting
- assessment of operatives' footwear.

Improvements in manual handling

The Principal Contractor introduced the following:

- mechanical lifting devices available to all contractors on the site, co-ordinated by the Principal Contractor, maintained and certified by him as necessary
- overall assessment of what materials were needed to be delivered to site, where they needed to be used, where they could be stored, etc. This entailed requesting from all contractors plans of work and scheduled deliveries of materials, etc.
- inclusion within site induction training of a section on manual handling, hazards and risks and aspects of manual handling that have been identified within the site, using a training video
- requirement for all contractors to provide Risk Assessments and Method Statements for manual handling activities
- requirement for individual contractor tool-box talks and evidence of training to individuals

- alterations to specifications of materials so that smaller and less heavy sizes were delivered to site
- poster campaign around the site.

Painting and decorating

The Principal Contractor was aware that the majority of painters and decorators on the site were small companies or self-employed individuals. Their general awareness of health and safety was poor and other trades were complaining to the Site Agent. The Principal Contractor decided to implement an awareness campaign on health and safety for the painters and decorators and introduced the following:

- tool-box talks by the Principal Contractor's Site Safety Office on
  - working at heights
  - working with hazardous substances
  - working with power tools
  - working in close proximity to other trades and people
  - housekeeping
- requirement for Method Statements and Risk Assessment for all work activity
- easy access to mobile tower scaffolds provided by the Principal Contractor
- leaflet campaign on contact dermatitis of hands and arms
- early morning work co-ordination meetings to agree who would be working where, what the hazards would be, etc.

Improving health and safety awareness of site foremen and supervisors

The Principal Contractor included a section in the Construction Phase Health and Safety Plan to improve health and safety awareness of all contractor and sub-contractor site foremen and supervisors. During the tendering process, each contractor had to agree to sign up and commit to the site foreman initiative.

The Principal Contractor then introduced the following:

- special training sessions for all site foremen on the roles and responsibilities for managing health and safety
- regular site safety inspections carried out with each site foreman
- a safety review meeting held weekly and attended by all site foremen and supervisors
- accident prevention campaign which highlighted a safety topic every month and reviewed incidents and accidents
- guidance and best practice booklets produced by the Principal Contractor's own safety department for distribution to each site foreman
- site-specific safety audit checklists which each site foreman was required to complete and return to the Principal Contractor
- introduction of a 'Safety Default Notice' system which recorded poor standards of health and safety against each contractor.

Site safety campaign

The Client on a major retail store development project was the instigator of a site safety campaign which was operated by the Principal Contractor on the Client's behalf. The best performing contractor in respect of health and safety was awarded a monetary sum for donation to a charity of their choice.

The Principal Contractor decided that the award should be given to the contractor who had made the most and greatest contribution to overall site safety. This involved not only ensuring that their own operatives were highly safety conscious, but also that they conducted their undertaking in such a way that the safety of others was improved.

The Principal Contractor arranged information meetings with all the contractors on site and set out clearly the performance criteria expected.

Assessments should be made on the following:

- frequency and effectiveness of tool-box talks
- practical, working knowledge of operatives on site on general health and safety principles
- quality and effectiveness of Method Statements and Risk Assessments
- co-operation and co-ordination of work with other contractors
- general commitment to health and safety, demonstrated by pro-active attendance at the weekly site safety meeting.

The Principal Contractor provided the services of the on-site Safety Office to any contractor who wished to have extra assistance in upgrading their health and safety practices. The Principal Contractor provided all contractors with a basic training pack for tool-box talks and attended many as an observer. Contractors were required to provide evidence of tool-box talks and individual operatives were interviewed to establish their learning outcomes.

The campaign ran for a three-month period and the Principal Contractor and the Client, assisted by the Planning Supervisor, assessed all information, accident records, etc. and decided on an outright winner.

A presentation award ceremony was arranged in the site canteen and the Client's Development Director presented a cheque for £500 to the successful contractor. In turn, they presented a cheque to a children's disability charity.

Similarly, on a major pharmaceutical company's research and development campus construction site, the Managing Contractor, as Principal Contractor, implemented an 'Accident Free Working Hours' campaign and set an objective to reach a million working hours free of any major accident or incident. The Site Safety Officer ran numerous campaigns on the site on safety topics, supported by poster campaigns, training, site audits, feedback to contractors, etc.

When milestones of accident-free working hours were met, all operatives on the site were awarded some tokens of achievement such as teeshirts, mugs, kit bags, etc. When a major milestone such as 500 000 accident-free hours was achieved, a major 'reward' was issued such as fleece jackets.

Although the campaign in itself might have been an additional cost on the project, the benefit was immense because the entire project lost no down time for accident investigations, stoppages, poor operative performance, etc. The project came in on time and predominantly on its half billion pound budget!

References

Management of Health and Safety at Work Regulations 1999.

Approved Code of Practice — Management of Health and Safety at Work Regulations 1999: L21.

Successful Health and Safety Management: HSG 65.

Managing Risks: Adding Value: HSE Books.

Health and Safety in Construction: HSG 150.

Managing Health and Safety in Construction: HSG 224.

5

Site welfare facilities

Is there a legal duty to provide welfare facilities on a construction site?

Yes. The Construction (Health, Safety and Welfare) Regulations 1996 contain specific regulations and accompanying schedules for the provision of welfare facilities.

Generally, everyone who works on a construction site must have access to:

- sanitary accommodation
- washing facilities
- warming facilities
- somewhere to eat their food
- somewhere to store clothes
- drinking water.

The Regulations tend to be 'goal setting' which means they do not stipulate exact numbers of facilities for every site. 'Suitable and sufficient' is a term which is often used and the employer has to determine what this might be.

How is 'suitable and sufficient' or 'reasonable' determined?

To an extent, common sense should prevail.

There must be enough facilities for everyone to use them without excessive waiting, etc.

Guidance on appropriate numbers is provided by the Health and Safety Executive or within other documents, e.g. British Standards. Employers are expected to know about the existence of other guidance and would be expected to consult good practice guides.

Should a Construction Inspector from the Health and Safety Executive visit site and declare that facilities are inadequate or not suitable and sufficient, an Improvement Notice could be served requiring the provision of additional facilities.

If the employer feels that the Inspector is being unreasonable, he could appeal the Improvement Notice and the matter would be heard at a Tribunal. Suitable and sufficient could be determined in this arena, or ultimately, in a Court of Law.

Do facilities have to be provided on the site or could toilets down the road be used?

In the majority of instances, welfare facilities will be expected to be provided actually on the construction site. It will be unreasonable to expect operatives to walk off site to find public conveniences, etc.

However, it may be reasonable to make proper arrangements with another employer to use their facilities elsewhere in the building for instance.

The HSE Inspector would expect to see a proper arrangement covered in the Construction Phase Health and Safety Plan (if the CDM Regulations apply to the project) or within information given to employees.

How many water closets, urinals and wash hand-basins have to be provided?

The Construction (Health, Safety and Welfare) Regulations 1996 state the facilities should be 'suitable and sufficient'.

Guidance on actual numbers of facilities can be found in other documents, namely the Approved Code of Practice for the Workplace (Health, Safety and Welfare) Regulations 1992 and British Standard 6465.

A good starting point for calculating the number of facilities is to use the following table:

No. of men at work	No. of water closets	No. of urinals	No. of wash stations
1 to 15	1	1	2
16 to 30	2	1	3
31 to 45	2	2	4
46 to 60	3	2	5
61 to 75	3	3	6
76 to 90	4	3	7
91 to 100	4	4	8
Above 100	An additional WC for every 50 (or part) men plus an equal number of additional urinals, plus an additional wash-hand station for every 20 operatives		
	Wash-hand stations should be provided with adequate supplies of hot and cold running water. Water closets should preferably be wash down water types		

Do toilets always have to be flushed by water?

No, not always, but it is preferable to have flushing toilets if at all possible.

If water supply and drainage cannot be provided to the site welfare facilities it will be acceptable to provide chemical closets.

Suitable numbers of chemical closets must be provided and they must have suitable mechanisms for maintaining the closets in a sanitary condition.

The Site Agent must ensure that greater attention is given to ensuring that chemical closets are kept clean and regular cleaning schedules must be in place. Regular emptying of the sewage containers will be necessary and plans will need to be made to deal with this.

If drains are available for discharging chemical closets the question might be asked as to why flushable toilets cannot be used.

What hand washing facilities would be acceptable if no running water is available?

Suitable facilities for washing hands and arms are essential on a construction site because the risk of hand to mouth infection is high.

If no running water is available, suitable containers of water must be provided, e.g. plastic containers with taps usually associated with camping and caravanning.

If hot water cannot be provided it is important to provide anti-bacterial soaps which work in cold waters, or water-less hand gel which effectively sanitises the hands.

Cement dust particularly needs to be removed from hands and arms so as to prevent skin diseases.

Do urinals have to be provided?

Not necessarily, as the overall number and accessibility of facilities is the most important factor. Urinals can be provided in addition to water closets and, where they are, a slightly lesser number of water closets will be needed.

Is it a legal requirement to provide showers on the construction site?

The Construction (Health, Safety and Welfare) Regulations 1996 are not specific about the need to provide showers for operatives working on a construction site.

The provision of showers should be considered as part of the Risk Assessment (or welfare facilities). They should be provided as a matter of good practice if operatives are to come into contact with particularly dusty activities or where there is a risk of skin contamination from substances. If work is particularly strenuous it may be reasonable to expect operatives to shower before leaving the site. The provision of good site welfare and washing facilities tends to set the standard of an overall well-managed site and the HSE would be pleased to see the inclusion of more showering facilities on construction sites.

Whose duty is it to calculate the number of sanitary facilities — the Client or the Principal Contractor?

Where construction projects fall under the CDM Regulations, the Client may stipulate what they expect to see in respect of site facilities. Clients are expected to set the standards for improving overall conditions on construction sites.

If a Client has set the standards, these will be found either in the Pre-tender Health and Safety Plan or in accompanying 'Employer's Requirements'.

The requirements listed in the Pre-tender Health and Safety Plan should be taken forward by the Principal Contractor and developed into the Construction Phase Health and Safety Plan.

If a Client has not stipulated any specific requirements, it will be assumed that the Principal Contractor will be responsible for ensuring that legal requirements in respect of facilities on the site are met.

The provision of site welfare facilities must be clearly covered in the Construction Phase Heath and Safety Plan. If the Client feels that insufficient provision is made he can prohibit start of the works on site under Regulation 10 of the CDM as the Construction Phase Health and Safety Plan will not be considered sufficient.

The Client should give an indication to the Principal Contractor of the anticipated number of contractors or workers expected on the site. This should help the Contractor to calculate numbers of facilities.

If site welfare facilities are shared between all contractors, who is responsible for keeping them clean?

The Construction Phase Health and Safety Plan (for CDM projects) should detail who is responsible for providing and maintaining welfare facilities.

Generally, this will be the Principal Contractor although another contractor could be identified as being responsible.

On CDM projects, the Principal Contractor is responsible for ensuring the co-ordination and co-operation of employers on a multi-occupied site.

Where facilities are shared with a residual employer, e.g. refurbishment projects in occupied buildings, agreement must be reached between both sides regarding who is responsible for cleaning and maintenance.

If facilities are found to be in an unsatisfactory condition, not clean, etc. and an HSE Inspector visits site, it is likely that an Improvement Notice will be served under the Health and Safety at Work Etc. Act 1974.

Is it necessary to provide separate sanitary and washing facilities for women?

Men and women may use the same toilet provided it is in a separate room with a door which can be locked from the inside. Where possible, cubicle walls and door should be full height, floor to ceiling so that the cubicle is totally enclosed.

Wash-hand basins can be shared between the sexes for hand and arm washing. It would be good practice to have a water closet and wash-hand basin in one cubicle but if this is not practicable, communal wash-hand basins in an ante-room to the water closets would suffice.

What provision needs to be made for Clients and visitors?

There is no legal requirement for separate facilities to be provided for Clients and visitors although many sites do have separate facilities.

A Client may stipulate that separate facilities are required and this will be a matter of agreement between contractor and Client. The HSE Inspector will only be concerned with the number of welfare facilities on site for the number of operatives working on the site.

Often, separate facilities are provided because they can be locked shut and more easily kept in an acceptable condition.

What provision needs to be made on a construction site for drinking water?

Schedule 6 of the Construction (Health, Safety and Welfare) Regulations 1996 requires that a suitable supply of drinking water be provided.

Drinking water is 'potable' water and meets the requirements of the Drinking Water Regulations. It generally needs to be a mains piped supply but adequate quantities of bottled water or water containers and dispensers will be satisfactory.

Every supply of drinking water shall be conspicuously marked by an appropriate sign where necessary for health and safety reasons. This is especially important where there may be two or more supplies of water around the site and one is fit for drinking while the others are not.

Where a supply of drinking water is provided, there must also be provided a sufficient number of cups or other drinking vessels unless the water supply is from a purposely designed drinking fountain.

On large sites, a suitable supply of drinking water must be provided at readily accessible and suitable places. This could be within each floor if the construction site is a multi-storey building or within, say, every 100–200 m.

It would *not* be considered reasonably accessible for only one drinking water supply to be available in the canteen if several floors need to be climbed to get to the facility.

More drinking water points need to be provided in warmer weather than in winter. If the site has a 'no food and no drink on site' rule it is imperative that the Principal Contractor makes adequate provision for drinking water supplies.

What facilities are required for the keeping of clothing on site?

Regulation 22 of the Construction (Health, Safety and Welfare) Regulations 1996 requires that 'suitable and sufficient' accommodation be provided or made available:

- for clothing of any person at work on the construction site and which is not worn during working hours and
- for special clothing which is worn by any person at work on the construction site but which is not taken home.

The facilities provided for keeping clothing on site must include suitable provision for drying clothing.

All operatives on a construction site have the right to keep their clothes safe during working hours, especially if they need to wear protective clothing which they put on when they reach site.

'Adequate' or 'suitable and sufficient' may include:

- coat hooks in the mess room
- coat hooks in an area adjacent to the sanitary accommodation
- a purposely designated hut with coat hooks, rails, etc.
- individual lockers.

Coat hooks, lockers, etc. could be shared.

Drying facilities will mean a room or an area where wet clothes can be hung to dry. A suitable heating device will be needed but this should be safe to use (e.g. avoid LPG heaters, open flames, etc.). Electrical heaters or drying rails will be ideal. Ensure that they are fitted with a safety, overheating cut-off device.

Protective clothing and equipment must be kept safe so as to keep it suitable for use. Separate storage facilities may be necessary.

Is it necessary to provide accommodation for changing clothes on site?

All construction sites must provide suitable and sufficient accommodation for the changing of clothes where:

- persons have to wear special clothing for work and
- that person can not be expected for reasons of proprietary or health, to change elsewhere.

It will be necessary to provide separate changing facilities for men and women working on the construction site.

Facilities could be a dedicated changing cabin, perhaps an integral part of the accommodation for storing clothes.

The ante-room in which wash-hand basins are located could double up as the changing room, provided it is for use by single sexes.

On large sites a locker room and changing room are often combined.

Personal possessions of operatives should be able to be stored on site securely — lockable lockers are preferred. Alternatively, personal possessions could be kept in a secure site office.

What is meant by the term 'rest facilities'?

All employees, self-employed and others 'at work' are entitled to have rest breaks from work. In fact, the Working Time Regulations require categories of employees to be given formal rest periods during the working day. These usually include tea/coffee breaks, lunch break and a mid-afternoon break. It depends on how long people are working before they are allowed a break.

If employees are expected to work for more than 6 hours continuously, they are entitled to a minimum of a 20 minute break.

Rest facilities cannot be part of the construction site area. They must either be a designated room or area. 'Rest' is to be a period of recuperation from noise, dust, activities, etc.

Operatives can eat their food and drink in rest areas. These should be away from the risk of contamination of the foodstuffs.

On smaller sites, if the site all stops at the same time and is cleared, an area within the construction site could be used as a rest area.

Suitable seating facilities should be provided and preferably a table.

Rest rooms and rest areas should be made available to separate those who smoke and those who do not. Passive smoking is a health hazard and adequate precautions are required to protect non-smokers from tobacco smoke.

Rest rooms must also contain suitable provision for the eating of meals, including the preparation of meals. A microwave oven would be suitable, together with table and chairs.

Some means of boiling water must be provided — again, a microwave oven would suffice. Operatives must have the opportunity to make hot drinks if required.

Rest rooms or areas must be adequately ventilated. They should not be exposed to risks from dust, noise, trailing cables, etc.

Rest rooms or areas should not be used to store plant, equipment or materials.

HSE Inspectors will expect to see adequate provision made on *all* sites for rest facilities. The details should be included in the Construction Phase Health and Safety Plan (where applicable).

Consideration needs to be given in *advance* as to whether the rest facilities need to be moved because of the progress of works on site. Plan in advance and always ensure that some defined area is available.

What provision for heating has to be made on a construction site?

Indoor working temperatures have to be reasonable — there is no maximum or minimum temperature set down in the Construction (Health, Safety and Welfare) Regulations 1996 (nor indeed in other health and safety legislation, although there is guidance).

Reasonable working temperatures are a matter of interpretation of a number of circumstances:

- What work needs to be done — how physical is it?
- How open is the site to the elements?
- How many people are on site?

- How easy is it for operatives to leave the work area to get warm?
- How much thermal clothing can be worn comfortably?

Generally, in order to arrive at a suitable decision as to what level of heating is required, a Risk Assessment will be necessary. The Principal Contractor should complete the Risk Assessment. The hazards from too cold a working environment are:

- increased risk of accidents
- lack of concentration
- increased risk of heart attack
- hypothermia
- poor circulation of the blood.

Obviously, the biggest concern is the likely increase in accidents if people are too cold to hold tools effectively, mix materials, etc.

The Principal Contractor must identify what control measures can be put in place to reduce the hazards and risks associated with low working temperatures.

Space heating could be installed in suitable locations throughout the site. Heating appliances are:

- electric
- liquid petroleum gas (LPG)
- gas.

Inadequately ventilated LPG and gas heaters could cause carbon monoxide gas to be produced and this could lead to fatalities. Gas equipment may continually leak because valves have been left on. These potential hazards need to be weighed up when choosing a suitable heating source.

Electrical fan blowers purposely designed for construction sites and running on 110 volts may be the safest option.

What provision for ventilation has to be made for a construction site?

All workplaces must be provided with adequate ventilation which means a supply of purified air or fresh air. Construction sites are no exception although, generally, there is little difficulty in providing adequate ventilation due to the open nature of many sites.

Particular attention has to be paid to the adequacy of ventilation when dust and chemicals, fumes and vapours are produced around the site.

Extract ventilation may be needed at certain times to eliminate dust or fumes. This requirement should be covered in the COSHH Assessment.

It would be sensible to complete a Risk Assessment to determine the needs for ventilation. Certain areas of the site may be 'confined spaces' and additional provision may be required.

Increased ventilation may be needed in summer months — a constant review of conditions on site is necessary so that adaptations to ventilation requirements can be made.

What provision needs to be made for lighting on the site?

As expected, the term for lighting on site is 'suitable and sufficient'. It should also be, where practicable, by natural light.

If artificial lighting is used it must not cause any warning signs or symbols to be adversely affected by a change of colour, etc.

Lighting must be suitable and sufficient and the only test is 'can you see where you are going around site?'

Is it possible to see the floor clearly, any small holes, drainage channels, etc. Can the task at hand be seen clearly without eye strain? Can operatives see what they are doing? Can they see the emergency exit signage? Can they see clearly when using steps and stairs or are they at risk of falling? Is there a difference between outdoor sunlight and internal lighting which may cause temporary blindness?

Is artificial lighting in the right place? Does it have enough lux? Poor lighting increases the risk of accidents. Improve the lighting levels and the accident rate may reduce, causing the site to be much more efficient.

Early morning starts and dusk in winter need to be considered and more lighting will be necessary.

Is it clear what lighting is to be provided by whom? It should be clearly laid out in the Construction Phase Health and Safety Plan. The Principal Contractor is responsible for background lighting and for lighting means of access and egress to the site. Contractors and sub-contractors may be responsible for providing task lighting for their individual trades.

The Principal Contractor should discuss the aspect of lighting in pre-start meetings and should check the contractor's risk assessments.

The Principal Contractor will also be responsible for ensuring that adequate provision is made on the site for secondary lighting when it is considered there would be a risk to health and safety if the primary lighting failed.

Emergency lighting will be required on fire exit routes. The Fire Plan should deal with these issues.

References

Construction (Health Safety and Welfare) Regulations 1996:

- Regulation 22
- Regulation 23
- Regulation 24
- Regulation 25.

Provision of Welfare Facilities at Fixed Construction Sites. HSE Information Sheet No. 18 (revised).

Health and Safety in Construction: HSG 150 (second edition).

6

Access and egress to site

What are the duties of the Site Agent in respect of safe means of access to site?

Under Section 2 of the Health and Safety at Work Etc. Act 1974, an employer is responsible for ensuring that employees and others have safe means of access and egress to their place of work.

A 'place of work' can be anywhere where an individual is expected to perform their duties and can include buildings, rooms, open spaces, working platforms, roofs, scaffolding, etc.

The Site Agent has a duty on behalf of his employers to ensure that employees have safe means of access and egress. In addition, the Site Agent has to ensure that persons other than those in his employ have safe means of access and egress.

Safe means of access mean access without risk of injury or harm.

There should be no risk of being run over by vehicles, no risk of tripping or falling over materials, plant, etc., no risk of falling from any height.

Safe access needs to be considered in relation to whether tools and materials need to be carried to the place of work.

Under the Construction (Design and Management) Regulations 1994 the Principal Contractor has the duty to prevent unauthorised access to site.

Where CDM Regulations apply to the project, the Site Agent must include in the Construction Phase Health and Safety Plan the details in respect of safe access to site, e.g. the preferred pedestrian route, how and where materials and plant will be delivered, vehicular access, routes out of site, emergency exits, etc.

Can pedestrians and vehicles use the same access routes?

It is important to keep vehicles and pedestrians as separate as possible as a high number of accidents, including fatalities, occur when vehicles and pedestrians share access routes.

Regulation 15 of the Construction (Health, Safety and Welfare) Regulations 1996 requires that construction work be so organised that, where practicable, pedestrians and vehicles can move safely and without risks to health.

In particular, traffic routes will not be considered safe if:

- steps have not been taken to ensure that pedestrians, when using traffic routes, can do so without danger to their health and safety
- any door or gate opens directly into the traffic route without any clear view of approaching traffic or vehicles
- pedestrians cannot have a place of safety to view oncoming vehicles
- there is no adequate separation between pedestrians and vehicles.

So, pedestrians and vehicles can use the same traffic route but *strict* safety rules apply and, generally, a Risk Assessment will be necessary which identifies the hazards and risks.

What additional precautions can be taken to separate pedestrians and vehicles, or manage the risks to their health and safety?

A number of safety precautions can be taken to ensure the safety of pedestrians:

- designate a safe walking area by painting hatch markings on a coloured walkway
- put up guard-rails to delineate the walkway
- use a banksman to direct vehicles and pedestrians
- use mirrors so that drivers and pedestrians can see routes clearly
- ensure that all vehicles have audible warning devices — not only for when they are reversing
- have a policy for all vehicles to ensure that lights and hazard lights are on during all hours, not just dawn and dusk.

What will constitute a safe means of access to upper work levels?

Where practicable, permanent means of access to upper levels will be expected, e.g. installation of the permanent staircase is preferable.

If the permanent structure can not be installed, then a purposely designed temporary staircase is preferable.

When determining safe means of access, remember to consider what people have to carry to their place of work, e.g. materials and tools.

Ladders are not necessarily classed as a safe means of access, especially for longer term projects. Ladders give immediate access to higher levels — they are not necessarily safe.

Safe means of access to upper levels may be by way of hoists and lifts.

Mobile elevating platforms are safer for accessing high level works, e.g. ductwork, ceiling works, etc. They also provide a safe working platform.

What other precautions need to be considered for safe access and egress to the place of work?

Access routes must:

- be clearly lit at all times
- be free of obstructions and trailing cables
- have clearly defined steps or slopes
- have handrails if changes in level are significant
- be of adequate size for the number of operatives using them
- be protected form hazards
- not be underneath activities being carried out at height (e.g. under a crane sweep, etc.)
- on stable ground
- clearly visible
- be designed to be away from material delivery points, etc.
- be identified in the Pre-tender Health and Safety Plan
- consider any other access routes into the building, e.g. if the employer is still operating out of the building
- not create crush points during clocking on/off times
- be adequately signed so that people know where they are expected to go
- have designated crossing points if vehicle/traffic routes need to be crossed
- be highlighted on a plan at the entrance to the site.

Is it necessary to provide a security point and signing-in station?

It is good practice to have a site control point because it is necessary to know who is and is not on site at any one time for fire safety purposes.

As a minimum, a designated signing-in place is essential at the site entrance. A site log for the signing-in and out of site is essential. This log can act as a reference log for fire safety purposes and will help to check whether everyone is accounted for in the event of a fire.

On complex sites it is good practice to have a full-time security person to operate the control point. This allows not only strict control of persons on to and off site, but also allows for checks on deliveries, skip removals, etc. and may help the issue of materials and equipment thefts.

The Pre-tender Health and Safety Plan should indicate whether the Client expects the Principal Contractor to provide a full-time operated access control point.

The Principal Contractor has to take responsibility under the Construction (Design and Management) Regulations 1994 for preventing the access of unauthorised persons to site. A properly controlled access control point will discharge this responsibility and, even if someone should gain access to site and put themselves in danger, the Principal Contractor would have a defence.

There has been a great deal of development over the years in computerised security systems. These are based on a swipe card and computer database and record access times and exit times. Many systems also combine photo identity cards and are used for training record purposes.

Does the construction site have to have more than one exit point?

The number of exit routes from a construction site depends on the number of operatives working on the site and the size and complexity of the site.

The Construction (Health, Safety and Welfare) Regulations 1996 stipulate that a suitable and sufficient number of exit routes must be provided which can be used in an emergency. Any person on the site must be able to reach a place of safety quickly and without hindrance.

It is always good practice to have more than one exit route. The access or way in can also double up as the exit route and often, this will be the most popular route, as it is well known. There should be alternative exits form places of work if the travel distance to an exit route is more than 45 m if a clear run, or more than 30 m if the route is less direct.

On large sites, several exits from the place of work will be needed.

An exit route cannot be counted if it leads people back into the building or site or leads them to a dead end. An exit route must lead to a place of safety.

Do emergency exit routes have to have emergency lighting?

If emergency exit routes are to be used in poor daylight conditions, e.g. dawn and dusk, or daylight is generally poor (as in winter) emergency lighting will be essential on construction site emergency exit routes. Emergency lighting should come on if any artificial lighting fails or if visibility is low. Such emergency lighting must be automatic.

Emergency lighting is particularly essential on staircases leading out of the building — poor lighting could lead to people falling on the stairs causing potential bottlenecks and crushing hazards.

Emergency signs should also be clearly visible and the emergency fire signs should be illuminated. At the very least, photo-illuminescent signs can be used.

For more information see the Chapter 14.

References

Construction (Health, Safety and Welfare) Regulations 1996:

- Regulation 15
- Regulation 19.

Health and Safety in Construction: HSG 150 (second edition).

7

Vehicles and transport

Vehicles and mobile plant are commonplace on construction sites. What are the main hazards I need to be aware of?

First, that vehicles, mobile plant and pedestrians do not go well together and that vehicles and mobile plant are the causes of many accidents, many with fatal consequences.

The key hazards when using vehicles or mobile plant are:

- moving vehicles running over operatives
- overturning vehicles or mobile plant
- reversing vehicles
- vehicles or plant too close to excavation edges
- vehicles or plant positioned on unstable ground
- vehicles or plant coming into contact with overhead power lines
- vehicles or plant coming into contact with buried services
- restricted access to site
- restricted vision of vehicle and plant operatives
- untrained operatives driving vehicles
- leaving vehicles unattended, with keys in the ignition
- overloading and therefore overbalancing of mobile plant
- inadequate operating space when using vehicles or mobile plant.

While vehicles and mobile plant create hazards on the site, they are necessary. What do I need to do to reduce the hazards and risks to manageable levels?

Good site planning will help reduce hazards on the site. This could start at the design stage of the project and should involve the Planning Supervisor, as appointed under the CDM Regulations, so that all Designers on the project are aware of what vehicles and plant may be needed on the site.

Provide safe entry and exit points with adequate turning space and good visibility for drivers.

Ensure that there is good lighting and visibility in areas close to pedestrians. Provide additional 'street lights', avoid blind corners and obstructions. Keep a good sight line for drivers of vehicles.

Plan to keep vehicles and pedestrians separate by having different entrances and exits for vehicles and people. Make sure each entrance and exit is properly signed, with clear text or pictogram signs which can be seen from all areas.

Provide separate, barriered walkways for pedestrians.

Where vehicles in particular have to be used in close proximity to pedestrians, provide a banksman.

Is there anything extra I need to do for reversing vehicles?

More accidents are caused by reversing vehicles than any others and the safety record of any site can be significantly improved by managing reversing vehicles.

Consider a one-way system for the site. This should be done at Planning Stage and the Planning Supervisor needs to co-ordinate with the Designers any alterations to the site layout so that a one-way system can be accommodated. It may be necessary to slightly relocate the position of a building so that adequate access is made available.

If vehicles need to reverse around the site ensure that they are fitted with audible reversing alarms and lights.

In any highly populated area, a banksman will wear a high visibility jacket and be properly trained in the tasks to be undertaken.

Prevent persons from crossing or moving in the area until the vehicle has completed its manoeuvres.

Are there any specific steps to take regarding routeways around the site?

It is sensible to set out clear and signed routeways across the site.

Avoid blind corners, sharp bends, narrow gaps and places with low headroom.

Avoid steep gradients, adverse cambers, shafts and excavations.

Provide a temporary road surface — this not only helps delineate the roadway but also helps to provide a level, stable surface on which vehicles can travel.

Introduce a regular inspection and maintenance programme for all of the route-ways. Potholes create hazards which could cause vehicles to overturn or loads to dislodge. Regular repairs need to be instigated.

Keep vehicles away from temporary structures, especially scaffolds, as an accidental knock of a scaffold pole could cause the scaffold to collapse.

Erect speed limit signs, e.g. 5 mph *maximum* speed limit throughout the site. Enforce the rules!

Erect directional signs to the entrance, exit or materials delivery area.

Reduce the amount of mud transferred around the site, and ultimately onto the highway, by installing wheel washers at key locations, e.g. entrances and exits.

Protect any excavations with barriers which both highlight the excavation and provide a guard to prevent vehicles falling into the void.

Do not allow other vehicles to park on route-ways, nor allow route-ways to be used for the storage of materials or plant.

What precautions are needed for vehicles which carry loads?

Make sure that vehicles which carry loads have been designed to carry loads — do not improvise and adapt vehicles.

Loads should be securely attached and any loose materials, e.g. bricks, timber, etc. must be secured with netting or tarpaulin or a similar covering. Materials blowing off vehicles create major health and safety hazards and falling material can cause serious injuries to site operatives and pedestrians.

Vehicles must not be overloaded as they will often become unstable. Vehicles used for carrying or lifting loads will have a Safe Working Load limit — make sure this limit is understood and weights are within the limits. Overloaded vehicles are also difficult to steer and their braking efficiency is impaired.

Operatives who drive vehicles must be over 18 years of age.

All operatives must be properly trained to drive vehicles.

Case Studies

A labourer riding as an unauthorised passenger on a dumper, fell and struck his head on the road. He died of head injuries.

An untrained labourer thought he could drive a dumper truck to collect some debris. The keys were left in the ignition. He drove it over uneven ground, lost control and turned it over on the edge of an excavation. He was crushed to death as it overturned.

A poorly maintained dumper overturned into an excavation causing the driver to be trapped under water. The braking system had failed due to lack of routine servicing.

Are there any special precautions in respect of health and safety to be taken when using dumper trucks?

Compact dumpers, the official name for site dumpers, are responsible for approximately one-third of all construction site transport accidents. The three main causes of accidents involving dumper trucks are:

- overturning on slopes and at the edges of excavations, embankments, etc.
- inadequately maintained braking systems
- driver error due to lack of experience and training, e.g. failure to apply the parking brake, switch off the engine and remove the keys before leaving the driver's seat.

The way to avoid such accidents is to pay attention to the following:

- ensure that all dumper trucks in use have 'roll-over protective structures' (ROPS) and seat restraints
- ensure that where there is a risk of drivers being hit by falling objects or materials, dumper trucks are fitted with 'falling object protective structures' (FOPS)
- (both ROPS and FOPS are legally required on all mobile equipment. If the equipment is hired, the hire company must ensure that the equipment complies with the law)
- ensure that a safe system of work operates when using dumper trucks
- provide method statements and risk assessments for using dumper trucks
- ensure that a thorough maintenance check is carried out and, in particular, that the braking system is checked
- check drivers' training records, driving licence and general competency to drive vehicles
- operate a strict no alcohol or drugs policy and prohibit anyone from driving who appears unfit to do the job

- plan any additional precautions necessary for using dumper trucks in inclement weather — restrict their use in icy conditions
- provide adequate stop blocks to prevent dumper trucks falling into excavations, etc. when tipping.

What are the key safety requirements for operating fork-lift trucks safely?

Every year there are about 8000 accidents involving fork-lift trucks (HSE Information) and, on average, ten of them involve fatalities.

There are a few simple measures which can be taken to manage the use of fork-lift trucks on a construction site, namely:

- managing fork-lift truck operations using safe systems of work
- providing adequate training for operators, supervisors and managers
- using suitable equipment for the job to be done
- laying out premises in such a way as to ensure that fork-lift trucks can move safely around
- ensuring that fork-lift trucks are maintained safely
- ensuring that the premises and site in which they are to be used are maintained in safe condition, e.g. potholes in access roads infilled, etc.

What legislation applies to the use of fork-lift trucks?

The main legislation includes:

- Health and Safety at Work Etc. Act 1974
- The Management of Health and Safety at Work Regulations 1999

- The Provision and Use of Work Equipment Regulations 1998
- The Lifting Operations and Lifting Equipment Regulations 1998
- The Workplace (Health, Safety and Welfare) Regulations 1992
- The Construction (Health, Safety and Welfare) Regulations 1996.

Breaches of sections of the Health and Safety at Work Etc. Act 1974 can incur fines of up to £20 000 and for the other Regulations fines are up to £5000 per offence.

How do I know if a fork-lift truck operator is properly trained?

All fork-lift truck drivers and operators must be able to demonstrate that they have received proper training from a competent instructor and from a recognised training body.

Operatives should have a valid certificate of training issued by one of the following:

- Association of Industrial Truck Trainers
- Construction Industry Training Board
- Lantra National Training Organisation Ltd
- National Plant Operators Registration Scheme
- RTITB Ltd.

Operators should be able to show the three stages of training as being completed:

- basic
- specific job
- familiarisation.

Basic training should cover the skills and knowledge required to operate a fork-lift truck safely and efficiently.

Job training should be site-specific and should include:

• knowledge of the operating principles and controls of the fork-lift truck to be used especially where these relate to handling attachments specific to the job
• knowledge of any differing controls on the machine to be used as this may be different to the one they were trained on
• routine servicing and maintenance of the fork-lift truck in accordance with the operator's handbook as may be required to be carried out by the operator, e.g. pre-start safety checks, visual checks, oil and brake checks, etc.
• information on specific site conditions, e.g. slopes, overhead cables/beams, excavations, one-way vehicle routes, confined spaces, designated exits and entrances, speed limits, site rules, etc.
• details of the tasks to be undertaken, type of loads to be carried, hazardous areas, materials, loading and unloading areas, etc.

Familiarisation training should take place on the site with the operator being supervised by a competent person. A full 'walk through' the site is recommended so that hazardous areas and site layout can be explained. Also, emergency procedures must be covered during site familiarisation.

Even when operators have formal certificated training qualifications in operating fork-lift trucks it is sensible to record on a site training record the subjects covered on the site familiarisation stage and have the operator sign acknowledgement.

Case Study

An employer had to pay out nearly £30 000 in fines and costs for an accident involving the operator of a fork-lift truck.

The employee who was driving the truck was crushed to death when it overturned.

A hole to lay the sewer pipes had been dug in the yard of a new building being constructed. On the day of the accident, the employee was using the fork-lift truck to load goods onto the back of a lorry. He reversed the fork-lift truck across the yard but one of the wheels slipped into the hole and the vehicle toppled over. The driver was trapped beneath the vehicle and suffered severe crushing injuries and died at the scene.

The employer failed to ensure that the surface of the yard was suitable for a fork-lift truck. Everyone knew about the hole but it was getting bigger due to water erosion. The employer had no system in place to monitor the condition of the yard and had not required a safe system of work, nor had he required Risk Assessments.

Prosecutions were brought under the Workplace (Health, Safety and Welfare) Regulations 1992, Regulation 12.

Top tips

Safe driving practices

- Check tyres, brakes and operating systems on all vehicles at the beginning of every day.
- Wear protective clothing and equipment, e.g. ear defenders and high visibility jackets.
- Use dumper trucks with ROPS and FOPS.
- Use vehicles with reversing alarms and adequate lighting.
- Follow site speed limits.
- Check that any loads are evenly distributed.
- Do not overload vehicles.
- Know the characteristics of the vehicle in all weather conditions.
- Ensure that all drivers are trained and competent.
- Separate pedestrians from vehicles.
- Do not stand on vehicles when they are being loaded or unloaded.
- Select neutral gear, switch off engine and remove keys when stopping and leaving any vehicle.
- Keep to designated route-ways.
- Make sure stop blocks are used.
- Do not use vehicles on steep inclines, adverse cambers, etc. without planning the job safely.
- Consider the unexpected and have a plan of action ready.
- Ensure good visibility at all times — do not overload dumpers, etc. so as to restrict vision.
- Check the area via mirrors, shouting, etc. before moving off.
- Do not drive any vehicle if unfit to do so for any reason.

References

Construction (Health, Safety and Welfare Regulations) 1996:

- Regulation 17

Workplace (Health, Safety and Welfare) Regulations 1992:

- Regulation 12

Health and Safety in Construction: HSG 150 (second edition).

Safe Use of Vehicles on Construction Sites: HSG 144.

Workplace Transport Safety: HSG 136.

Safety in Working with Lift Trucks: HSG 6.

8

Excavations

What are the legal requirements governing excavations?

Regulation 12 of the Construction (Health, Safety and Welfare) Regulations 1996 sets out the requirements for excavations.

First, however, Regulation 2 of the above Regulations must be consulted in order to define the term 'excavation'.

In Regulation 2 (Definitions) an excavation is deemed to include:

- earthworks
- trench
- well
- shaft
- tunnel
- underground working.

Regulation 12 sets out the following.

All practical steps shall be taken, where necessary, to prevent danger to any person, to ensure that new or existing excavation or any part of such excavation which may be in a temporary state of weakness or instability due to the carrying out of construction work including other excavation work) does not collapse accidentally.

Suitable and sufficient steps shall be taken to prevent, as far as reasonably practicable, any person from being buried or trapped by a fall or dislodgement of material.

Where it is necessary for the preventing of danger to any person from a fall or dislodgement of any material from the side or roof of or adjacent to any excavation, that excavation shall, as early as practicable in the course of work, be sufficiently supported to prevent, so far as is reasonably practicable, the fall or dislodgement of such material.

Suitable and sufficient equipment for the supporting of an excavation shall be provided to ensure that the requirements of the above sections may be complied with.

The installation, alteration, dismantling of any support for an excavation shall be carried out only under the supervision of a competent person.

Where necessary to prevent danger to any person, suitable and sufficient steps shall be taken to prevent any person, vehicle or plant and equipment, or any accumulation of earth or other material from falling into any excavation.

Where a collapse of an excavation would endanger any person, no material, vehicle or plant and equipment shall be placed or moved near any excavation where it is likely to cause such collapse.

No excavation work shall be carried out unless suitable and sufficient steps have been taken to identify and, so far as is reasonably practicable, prevent any risk of injury arising from any underground cable or other underground service.

What does the Regulation actually mean?

The Regulation states that excavations are dangerous and steps must be taken to prevent danger and harm to people.

Working in excavations requires a planned approach which identifies risk assessment, safe systems of work, monitoring and constant review of procedures.

The safety of all people associated with the excavation and those who may be in close proximity must be considered.

The Regulation does not define 'suitable and sufficient steps' and it is for the employer or person in control of the site to determine what needs to be done to comply with the 'suitable and sufficient' requirement.

The Regulation implies that a Risk Assessment should be carried out for excavation works. It is only when this has been done that the precautions necessary to prevent danger can be identified.

Once the hazards and risks have been identified in the Risk Assessment, control measures need to be implemented to reduce the risks to acceptable levels. The method statement will formulate the controls and 'system of work' which will need to be undertaken.

Is there any other legal duty with regard to excavations?

Yes. Excavations must be inspected by a competent person before work commences in them. Either the employer of the operatives or the person controlling the works on site — usually the Principal Contractor — must carry out the inspection.

The excavation must be inspected to confirm that it is 'stable and of sound construction' and that the safeguards required by Regulation 12 are in place.

Once the inspection has taken place, a report must be made of the findings. This report must be available for inspection by any person and kept at the place of work. The findings must be communicated to the people using the work area.

When must an inspection be carried out?

The guidance for inspections can be found in Schedule 7 of the Construction (Health, Safety and Welfare) Regulations 1996.

Excavations must be inspected:

- before any person carries out work at the start of every shift
- after any event likely to have affected the strength or stability of the excavation or any part of it
- after any accidental fall of rock or earth or other material.

What information must be included in the report?

A report of inspection must include the following particulars:

- name and address of the person on whose behalf the inspection was carried out
- location of the place of work or part of that place inspected (including any plant and equipment or materials, if any)
- date and time of the inspection
- details of any matter identified that could give rise to risk to the health and safety of any person
- details of any action taken as a result of any matter identified which could cause safety issues
- details of any further action considered necessary
- name and position of person making the report.

There is no formal format for a report but guidance and templates are suggested by the HSE. As long as the above information is included, the employer can design his own form.

What needs to be considered when planning work which involves excavations?

The key to safe working in excavations is planning. Planning starts with knowing what ground conditions exist on the site and obtaining information. Before digging or starting any excavation it is essential to consider and plan against the following:

- collapse of sides
- materials falling onto people working in the excavation
- people and vehicles falling into the excavation
- people being struck by plant
- undermining nearby structures
- contact with underground services
- access to the excavation
- fumes
- accidents to members of the public.

It is important to ensure that materials needed to protect excavations are readily available on site *before* works to excavations start. Materials required may include:

- props
- trench sheets
- baulks
- timber planks
- ladders
- guard railings
- signage.

The Risk Assessment should have identified what materials and equipment are necessary.

What needs to be done to prevent excavations collapsing?

Ground movement causes trench or excavation sides to collapse. Often, the weight of the removed spoil placed adjacent to the excavation imposes an increased loading which exacerbates instability of the ground. The ground type must first be identified. The first rule about excavation collapse is:

- do not believe that any excavation is safe from collapse.

Excavations in semi-rigid soils, rock, etc. may look safe and they may stand unsupported from 30 seconds to 30 days. But there is no knowing when the ground may move and the sides collapse.

Spoil heaps also move and can cover and infill the excavation or surrounding areas.

There is no legal minimum height or depth of an excavation which must be shored or propped — a depth of 2 m is often used as a guide for when excavations need to be propped. However, this is erroneous and an excavation of only 1.0 m may be unsafe and need to be propped.

Any excavation where there is a fall height into the void of 2 m or more must legally be guarded with edge protection.

Excavations are prevented from collapsing by proper shoring and propping of the side walls.

Sides to excavations could be made safe be battering them to a safe angle but, generally, a safer option is to shutter the sides with timber planks and props, trench sheets specially designed for the job or specialised proprietary propping systems.

The Risk Assessment must determine at what intervals the propping shall take place, e.g. continuous or at, say, 1 m intervals. If propping is not continuous there may be a danger of partial collapse.

Propping or shuttering of an excavation must be properly planned by a competent person — for large and complex excavations a Structural Engineer may be necessary to design the system because the timber props, panels, etc. will need to resist horizontal lateral loads and other loading.

Sheeting and vertical props must be sunk deep enough into the ground to ensure stability. Many accidents have happened because the trench or excavation shuttering has collapsed due to inadequate design and installation.

Consideration must also be given to the risk of water ingress into the excavation or waterlogging of adjacent ground as this will weaken trench or excavation sides and increase the likelihood of collapse.

Water itself exerts great pressure and could cause any shuttering installed to fail.

Case Studies

A ground worker was buried up to his neck as he worked in a trench of approximately 2.0 m depth. Trench sheets had been used but only at 900 mm intervals. They had not been properly secured. A heavy breaker was being used to remove boulders from the ground. The vibrations caused the shoring to give way and the trench collapsed, trapping the ground worker.

Four ground workers were buried alive when the excavation they were working in collapsed. The trench was approximately 4.0 m deep and because the operatives believed the ground to be semi-rigid as it was mudstone, they did not heed advice and shore up the sides of the excavation. With no warning of impending collapse, the trench side gave way, trapping all four workers. Three died but one survived with multiple fractures.

A labourer suffered multiple fractures to his head and upper body when the brick wall situated next to the 900 mm trench he was digging collapsed into the excavation as he had undermined the walls foundations without realising. Even though his trench depth was only approximately 1.0 m, the foundations of the wall were at a greater height.

Excavation collapses are also prevented if there is an 'exclusion zone' placed around the excavation as this will prevent vehicles and people walking too near the sides or on spoil heaps, etc. thereby undermining the stability.

What steps can be taken to prevent people or equipment and vehicles falling into excavations?

The easiest and most effective safety measure to take is to protect the opening of any excavation so as to prevent people and equipment falling into it.

Any excavation which could cause someone to fall 2.0 m or more *must* have adequate edge protection and guarding.

Guard-rails must be substantial and withstand impact loads if people fall against them. They must have a top rail height of approximately 1000 mm, a mid-rail at 450 mm and a toe board at least 150 mm high.

Ensure that an excavation is properly signed with hazard warning signs. Ensure barriers, etc. used to guard the void are highly visible.

Protect the excavation while operatives are working in it so as to prevent persons or equipment falling in and potentially crushing the operatives.

When work in an excavation is finished for the shift it might be safer to fully board over the hole so as to prevent any risk of persons falling. Consideration must obviously be given to span widths, etc so as not to create an even greater hazard from unsafe surfaces.

Keep all people, vehicles, plant and equipment away from excavations. Create an 'exclusion zone' around the excavation. Barrier off the excavation well beyond the edges. Erect hazard warning signs. Make sure that the excavation can be clearly identified. Consider any hazards of poor lighting, restricted access, traffic routes and walkways, etc.

Complete Risk Assessments for any tipping activity into the excavation and ensure that stop blocks are used to prevent vehicles overrunning. Consider whether hand bailing infill spoil, etc. will be safer and stop the lorry away from the excavation edge.

Remember health and safety is about *accident prevention* so introduce safety procedures over and above the bare minimum, e.g. excavations shallower than 2.0 m may need edge protection.

What are some of the other hazards which need to be identified?

Excavations often involve the use of excavators, diggers, etc. and, often, more accidents are caused because people are killed by the vehicle or plant than by falling into or being crushed by the excavation.

Keep workers separate from moving plant. Where this is not possible, employ a banksman to guide the excavator and to protect people. Not all operatives on a site will be familiar with the hazards associated with excavations.

Develop safe systems of work which manage when, how, where and by whom excavation work is undertaken and decide how such work may impinge on the safety of others.

Structures often become undermined because of excavation works and buildings or structures can collapse into the excavation or in close proximity to it.

Structural Engineers should be consulted when excavations are to take place adjacent to buildings or structures. Details of foundation depths, etc. should be given to the Principal Contractor via the Planning Supervisor and the Pre-tender Health and Safety Plan.

What are the hazards from underground services?

Every year a significant number of ground workers are injured through contact with underground services.

Underground services include:

- electricity mains
- gas mains and distribution pipes
- water mains
- telecommunications cables
- sewers and foul drainage systems.

All have the potential to cause fatal injuries both to those carrying out the excavation and others in the vicinity of the work. Other injuries can include severe electrical burns, multiple fractures, head injuries, breathing of toxic fumes, etc.

Information on underground services should be available *prior* to commencing any excavation. Investigations should be carried out to confirm the information, or to ascertain the information.

Check utilities service drawings and plans.

Look around for obvious signs on site, e.g. valve covers, road repairs, inspection covers, etc.

Use pipe and cable locators and mark the ground accordingly.

If there is any possibility that unexpected services will be encountered, follow safe digging practices:

- dig trial holes with hand tools
- use spades or shovels
- do not throw or spike spades or shovels into the ground, use a gentle foot pressure
- use any picks, forks, etc. with care to break up larger clumps of soil or stone
- try to excavate alongside the anticipated service pipe or cable so as to avoid a direct hit
- use an 'air digging' tool which will remove soil by air pressure and prevent impact damage to services
- assume all services are live until a proper disconnection and Permit to Work system indicates otherwise
- when buried services have been uncovered, wait for identification before proceeding

Case Study

An operative using a powered pneumatic breaker lost both his arms when he hit a live 11 kV mains cable. The cable had been identified within a general area and the operative set about digging the trench. He had dug down only approximately 0.5 m when he hit the mains cable.

The investigation showed that no safe system of work was in operation at the site. No detailed investigation had been done to locate the exact position of the cable, no safe digging techniques were being followed.

The employer of the operative was prosecuted for failing to have a safe system of work and so was the Principal Contractor. Other contraventions included:

- no Risk Assessment and
- no training records.

- support any disturbed pipes or cables
- do not use pipes, etc. as foot- or hand-holds to exit a trench
- label any services found.

Ensure that a comprehensive Emergency Plan and Procedure has been drawn up in case of an accident, explosion, fire, etc.

All places of work have to have a safe means of access and egress. What does this mean for excavations?

The bottom of an excavation is often the place of work for someone and a safe means of getting in and out is essential. Accidents happen because access ladders, etc. are not secured or there is no safe way of getting materials and equipment into the trench or excavation.

A ladder is probably the most common method of access into an excavation. Ladder safety therefore prevails and the ladder must be:

- at the correct angle
- secured top and bottom
- protruding approximately 1.0 m beyond the top of the excavation depth
- maintained in good condition.

Excavations of any substantial length should have more than one access and egress so that alternatives are available in emergencies. A good practice guide is approximately 6 m. It may be necessary to provide additional lifting equipment, e.g. a hoist, in order for materials and equipment to be safely lowered into the excavation.

Are there any other hazards which may need to be considered regarding excavations?

Excavations could also be classified as confined spaces, e.g. shafts, and it may therefore be possible for any of the following to apply:

- build up of toxic fumes
- build up of toxic gases
- lack of oxygen
- accumulation of hazardous substances, chemicals, etc. that are heavier than air.

Even if not a confined space, gases and toxic fumes can be problems in excavations.

Sewers and drainage systems may give off hydrogen sulphide gas which can be fatal. In tolerable quantities, hydrogen sulphide gas smells like bad eggs and, when this is smelled, the excavation should be evacuated, ventilated and the source of the gas identified. But in its most concentrated levels where it is breathed in, hydrogen sulphide gas is odourless — the silent killer for drainage workers.

If any work is being undertaken near sewers, drainage pipes, inspection chambers, etc., a safe system of work must be followed. Gas detection devices are essential as they monitor the concentration of various gases.

Oxygen deficiency can also be a safety hazard — often caused by oxygen being pulled out of the excavation by pressure differences, or because of chemical reactions for fumes etc.

Carbon monoxide could build up where there is insufficient oxygen and, again, carbon monoxide is a silent, odourless killer gas.

Fumes may penetrate excavations from nearby plant which is generating exhaust fumes. Diesel and petrol-operated machinery should not be used near excavations unless the exhaust fumes can be directed well away from the area, or the excavation can be supplied with additional forced air ventilation.

References

Construction (Health, Safety and Welfare) Regulations 1996:

- Regulation 12
- Regulation 29

- Regulation 30
- Schedule 7.

Health and Safety in Excavation: HSG 185.

Avoiding Danger from Underground Services: HSG 47.

Health and Safety in Construction: HSG 150 (revised).

9

Working at heights

What are the legal requirements for working at heights?

The Construction (Health, Safety and Welfare) Regulations 1996 cover the legal requirements for working at heights.

Regulation 6 covers the subject of falls and, generally, requires the following:

- suitable steps must be taken to prevent persons falling
- provisions to prevent falling will include the use of barriers, guard-rails and toe boards
- anyone who is likely to fall a distance of 2.0 m or more, or who will use access or egress routes from which they may fall more than 2.0 m shall be protected by means of a suitable guard-rail
- working platforms must be properly designed and of the correct width for safe working
- fall arrest equipment must be used where work is of short duration but persons can still fall 2.0 m or more
- ladders must not be used as a means of access or egress to a place of work unless it is reasonable to do so having regard to the nature of the work being carried out and the risks to safety arising from using the ladder

- ladders shall comply with specific requirements contained in Schedule 5 of the Regulations
- equipment used for accessing or working at heights shall be properly maintained
- scaffolding shall be erected by competent persons.

Why are the requirements for working at heights so strict?

More accidents involving fatalities and major injuries are caused by working at heights than any other construction activity. Most accidents which happen involving working at heights are preventable.

Usually when someone falls from a height the injuries are quite significant. Even when falling from a height of less than 2.0 m, serious injury or death can result.

The greater the possible severity of injury, the greater the level of control needed to reduce the risks.

What steps do I need to take to reduce the risks from working at heights?

First, an assessment needs to be carried out to determine what work needs to be carried out at heights and whether there are other ways to perform the tasks.

Working at height activities often include:

- roof works
- working off ladders
- working from platforms
- cleaning windows
- ductwork installations

- ceiling works
- guttering works
- working on atria
- cleaning works.

Can a continuous platform be erected so that everything is at the same level?

Can staircases be installed so as to avoid access to higher levels via ladders?

Should scaffolding be erected for roof works, facade and guttering works, etc.?

The first principle of risk assessment is to eliminate the hazard wherever possible.

Designers should be looking at how to design safe access and safe working areas to high level parts of buildings. CDM Design Risk Assessments should be reviewed — how do the designers expect access to be gained to high working areas and how do they propose to keep operatives safe.

Carry out a Risk Assessment for working at heights, including how to get to and from the place of work.

Eliminate the hazard where possible.

If this is not possible, the hazard should be substituted for a lesser hazard. This means, for instance, that if work is to be done from a ladder it would be a lesser hazard to substitute a mobile elevating work platform for the ladder as there is less risk (although a risk still exists) of falling off a platform.

Where a hazard cannot be eliminated, controls must be put in place to manage the consequences of the hazard and reduce the risk and severity of injury.

As with any hazard and risk associated with a work activity, everyone must receive information, instruction and training about the hazard and risk.

Remember, fall *prevention* is much better that fall protection.

People generally underestimate the risks of injury from a task, especially if they have been doing the same job frequently over long periods of time. People therefore, find it difficult to make their own judgement on what protection to take.

Sometimes, health and safety controls need to be prescriptive and not subjective or interpretative.

What are some of the simple every day controls I can instigate on my site?

The most effective control to implement is a rigorous compliance with the requirement to guard all leading edges, drops, voids, lift shafts, holes, roof edges, etc. with suitable guard-rails where there is a risk of anyone falling *any* distance. However, if there has to be a standard, it must be that anywhere from which people can fall 2.0 m or more *must* be guarded.

What are the requirements for guard-rails?

Guard-rails must be substantial and properly fixed so that they are secure. Falling against a guard-rail which gives way is no protection at all.

Guard-rails must:

- have a top rail at a height of approximately 900–1000 mm
- have an intermediate guard-rail at an approximate height of 450 mm
- have a toe board of approximately 150 mm in height
- be continuous around an opening
- be clearly visible
- not be breached by openings or missing rails, etc. (e.g. at lift or hoist openings).

Guard-rails must be inspected regularly by a competent person and any defects repaired immediately. Hazard warning signs must be displayed if necessary but any defect to a guard-rail renders the rail unsafe and so work should be prohibited in the area until the guard-rail has been repaired.

When would it be reasonable to use fall arrest systems?

Fall arrest systems are personal protective equipment which does not prevent someone from falling but, if were to fall, *arrests* their descent so that they do not fall all the way to the ground.
Fall arrest systems usually consist of:

- lanyard or rope
- harness for the body
- hooks and couplings to connect the restraint ropes, etc. to the body harness
- hooks or couplings to connect the lanyard or rope to the securing point (e.g. eye bolt, permanent rail, etc.).

Fall arrest systems should only be used when fall prevention is impossible, e.g. when erecting scaffolding.
The Risk Assessment should state whether fall arrest systems are suitable control measures to reduce the risk of injury when working at heights.
Fall arrest systems may be the only safety precaution available if work is of a short duration at heights.

What needs to be considered when using fall arrest systems?

Harnesses and lanyards are made of man-made fibres and, as such, are subject to degradation by sunlight, inclement weather, chemicals, etc.

It is important to carry out detailed daily checks of the equipment prior to use. A visual and tactile inspection is necessary — look and feel for faults in the lanyard or harness. Choose an area in good light.

If there is the slightest doubt about the condition of any part of the fall arrest system do not use it. Label it as 'defective' and put it aside for further inspection.

Faults can be identified by:

- discolouration
- tears
- fraying
- nicks
- grittiness
- rust on metal catches
- excessive scratching.

Harnesses do *not* prevent a fall. The person falling may not fall to the ground but will be suspended in mid-air or only fall to the next level down. There is a significant risk of injury even when using a fall arrest system because people could be harmed by impact to the body when the lanyard goes taut or if they strike against parts of the structure during the fall.

Consider the use of an energy-absorbing device fitted to energy-absorbing lanyards so as to reduce the risk of injury from impact loads.

Minimise the 'freefall' distance. Keep the anchor as high as possible so as to reduce fall distances.

Ensure that anchor points are suitable and have been checked by a competent person. Any anchor points to which a fall arrest system is attached constitute 'lifting equipment' under the Lifting Operations and Lifting Equipment Regulations 1998 and *must* be inspected by a competent person every six months.

Ensure that the lanyard is not too long so that it has a chance to arrest the fall before the person hits the ground.

Anyone who needs to attach themselves to a fall arrest system needs to be able to do so before they have entered the area from which they need to be protected.

Develop a system of running lines and second lanyards so that a person can unclip themselves and clip themselves on in a continuous process without being put at risk of falling.

Ensure that everyone wearing a harness and lanyard knows how to wear it properly, how to check it, adjust it, etc. Everyone should carry out their own visual pre-use checks and must themselves be satisfied about its conditions.

Never force someone to wear a harness or lanyard when they believe it to be defective.

What needs to be considered in respect of temporary suspended access cradles and platforms?

Temporary suspended access cradles and platforms are often used for window cleaning and external facade maintenance and cleaning. They are usually fixed at anchor points on the top of buildings and the cradle passed over the edge to 'hand down' the building facade.

Many accidents happen when using suspended access cradles, mostly due to:

- unsafe access to and from the cradle, e.g. stepping over a parapet wall
- insufficient or poorly secured counterweights and holding systems
- failure of the cradle platform or components such as pins, bolts, etc.
- failure of winches, climbing devices, safety gear and ropes as a result of poor maintenance
- poor erection and dismantling techniques.

Equipment must be chosen and installed by a competent person. A comprehensive Risk Assessment is required.

Consideration must be given to:

- type of weather
- wind speeds
- access into the cradle
- egress from the cradle
- safe working loads of cradles to include number of operatives and type and number of equipment to be used and therefore kept in the cradle
- protection of ropes or suspension cables over parapet walls, etc. so as to prevent friction and possible fraying
- risk of overturning
- likelihood of suspension cable collapse and an alternative backup cable, etc.
- guarding of the cradle
- operating instructions
- training and experience of the operatives.

An emergency plan must also have been devised so that if there is a failure of the system or an injury to operatives in the cradle, emergency measures can be taken for rescue, etc.

Access into the cradle should preferably be from ground level.

All access cradles should be raised when not in use so that unauthorised access, especially by children, is prevented. Power should be switched off and isolated.

Access cradles must display suitable safety signs and have Safe Working Loads (SWL) displayed. SWLs should be equated to understandable information (e.g. number of people as opposed to kg weights).

Are there any common safety rules I could follow for working at heights?

Committing the following to memory and always following them will help to reduce injuries from falls significantly:

- do not work at height unless it is absolutely essential
- make sure any working platform is secure and stable
- check that any working platform will support the weight of workers and the necessary equipment
- make sure any access equipment or working platform is stable and will not overturn
- do not erect equipment on uneven ground
- ensure that everyone has adequate working space — overcrowded and restricted work areas contribute to accidents
- foot any ladders or access towers or secure them to a stable structure
- provide guard-rails, barriers to all edges, openings, drops, etc.
- check everything frequently and do not take anything for granted
- if fall arrest systems are used, check their condition regularly.

What other hazards are associated with working at heights?

Falling through fragile materials, especially on roofs, is a common and frequent hazard on construction sites and in maintenance works.
Fragile materials can include:

- asbestos sheeting
- asbestos cement
- fibreglass
- plastics
- steel sheets
- glass sheets.

The above materials may become brittle over the years when exposed to sunlight. Steel may rust.

Any material could give way without warning and such an eventuality must always be considered.

Roof openings and roof lights are common hazards which contribute to people falling.

Roof lights often become brittle but they may look in perfect condition.

Structural Engineers should be consulted if there is any doubt about the ability of any structure to carry loads.

What safety procedures need to be followed for working on fragile materials?

Generally, the first principle of risk assessment is to eliminate the hazard. So do *not* work on fragile materials unless there is absolutely no alternative.

Fragile materials may be disguised by loose coverings, moss coverings, etc. so nothing should be taken for granted.

Work on fragile roofs has to be carefully planned to prevent falls through the roof.

Working platforms to spread the load are essential.

Purpose-built roof ladders are essential — these hook over the ridge tiles and should be secured at the base. A review should take place to ensure that any tiles or parts of the structure used to secure roof access ladders are stable and secure.

If there is no possibility of working off platforms, then fall arrest systems, running lines, etc. must be considered.

Safety nets are also a possibility, although they do not stop people falling, but may reduce the possibility of serious injury.

Where there are localised areas of fragile materials, e.g. roof lights, these can be protected by adequate guard-rails.

Risk Assessments *must* be completed for all works undertaken at heights.

Other safety considerations are:

- access to the place of work
- weather conditions

- wind conditions
- methods of getting materials to the place of work
- falling materials and objects
- emergency procedures and rescue plans.

What are the common safety procedures for using scaffolding?

All scaffolds, including mobile tower scaffolds, should be designed, erected, altered and dismantled by competent persons. All scaffolding work must be supervised by a trained and competent person.

Scaffolders must adopt safe systems of work when erecting scaffolds and must wear fall arrest equipment.

Scaffolds must be adequately braced to prevent them from collapsing. Consideration must also be given to wind loadings and materials and weight loadings.

Scaffold walkways should be at least four boards wide.

All scaffolds must be sheeted or netted to prevent objects falling.

Hoists, chutes and lifting tackle must be assessed to establish the effect it may have on the scaffold.

Do not take up boards, move handrails or remove tiles to gain access to work areas.

Changes to scaffolds should only be made when properly assessed, planned and undertaken by competent people.

Scaffold lifts must never be partially boarded. Boards must be secured or must overlap each other by at least 150 mm. Consideration must be given to trip hazards on the walkway.

Boards forming the walkways must be adequately supported by horizontal transoms. Spacings should be between 1.2 and 1.5 m. Inadequate support will cause the boards to bow and give way.

Safe means of access must be provided by way of secured vertical ladders or ladders placed at a suitable angle for ease of use. All ladders must be tied.

Case Studies

A maintenance worker, aged 38, plunged 30 ft to his death after falling through an asbestos roof, an inquest heard. He went on to the roof of a building to repair panels which had been damaged in a break-in, after being told by his boss to 'sort it'.

A scaffolding company was fined heavily after a court heard how a scaffolder was killed when he fell from a third storey window-ledge to the street below.

The scaffolder was removing a large hoarding from the front of a building. The scaffolder climbed through the window on the third floor in order to gain access to a narrow ledge so that he could stand on it to reach the hoarding brackets.

He lost his balance and fell to his death.

The Company was prosecuted for:

- failing to have a safe system of work
- failure to carry out Risk Assessments
- failure to provide employees with adequate information, instruction and training.

The HSE Inspector said that a mobile elevating work platform should have been used. The procedure adopted was inherently unsafe.

A Store Assistant who was working in a retail store died from head injuries after falling from a ladder, an inquest heard.

He was using the ladder to remove a bike part from an upper shelf in the storeroom when he fell from it.

A colleague found him on the concrete floor on all fours with blood coming from his head.

'He was across the ladder with a cycle against him. He was bleeding from the head, he was dizzy, delirious and started being sick. I rang for an ambulance immediately,' he said.

The Store Assistant underwent brain surgery in a local hospital but died four days later.

The inquest jury returned a verdict of accidental death.

The Health and Safety Inspector said he had investigated the scene and the ladder being used lacked proper feet and had been used at an incorrect angle.

His injuries were severe and were described in court:

'He had an abrasion and swelling on his head and there was fresh blood leaking from his right ear. He had a brain scan and suffered a fit while in the machine. The results of the scan showed serious head injuries, two skull fractures, bleeding around brain tissue and significant bruising in two parts of the brain.'

Scaffolding must be inspected:

- before first use (i.e. after erection)
- after substantial alteration
- daily before use
- after any event likely to have affected stability (e.g. high winds)
- at regular intervals not exceeding every seven days.

Suitable records of inspection shall be kept.

Faults must be highlighted clearly and the scaffold put out of use until repaired — the only sensible action to be taken for a defective scaffold.

Any contractor using a 'communal' scaffold has a duty to inspect the scaffold and to be confident about its condition *before* any of his operators use the scaffold.

References

Construction (Health, Safety and Welfare) Regulations 1996:

- Regulation 6
- Regulation 7
- Regulation 8
- Schedule 1
- Schedule 2
- Schedule 3
- Schedule 4
- Schedule 5.

Health and Safety in Roof Work: HSG 33.

Working on Roofs: HSE INDG 284.

10

Electrical safety

I understand the law dictates that 230 v electricity cannot be used on a construction site. Why?

Strictly speaking the law does *not* state that you cannot use 230 v electrical supply on a construction site. You could do so if you manage the risks adequately.

However, 230 v electricity is a major hazard and has great potential to kill and cause injury to people if used unsafely or in conditions which are unsafe.

Managing health and safety is about eliminating or reducing hazards to acceptable levels.

Requiring contractors to use 110 v power tools or 12 v power tools significantly reduces the likelihood of injury from the electrical supply.

It is therefore a safe system of work to use 110 v or 12 v power tools.

What safety steps must I take if 230 v is the only available power on site?

If power tools have to operate on 230 v then the following are required as an absolute minimum:

- detailed Risk Assessment
- all equipment, cables and leads must be in good repair
- residual current devices (or trip devices) which operate at 30MA with no time delays must be used
- regular daily checks must be carried out
- hazardous environments must be avoided (e.g. eliminate damp conditions, water, dust, use of chemicals, etc.).

All equipment must be connected to RCDs or, if agreed by a competent electrician, an RCD can be fitted to the main power supply so that it protects the whole of the electrical system.

Risk Assessments must be constantly reviewed and updated should conditions on site change.

Doubly insulated equipment should be used where possible.

Residual current devices must be installed with great caution as a poorly installed one will create more hazards than it solves.

RCDs must be:

- kept free of moisture
- kept free of dirt
- protected from vibration
- protected from mechanical damage
- properly installed and enclosed
- properly protected with sealed cable entries
- checked daily by using the test button
- considered as a safety device and not the complete eliminator of a hazard.

Tools must only be connected to sockets protected by RCDs.

Cables and leads supplying equipment running on 230 v must:

- be protected from damage
- kept at high levels
- protected inside impact-resistant conduit.

For additional safety, use armour-protected flexible cables and leads.

What should the Site Agent do if a contractor comes on to site and wants to use 230 v equipment when the site operates on 110 v?

The first response will be to refuse him access to the site on the grounds that he will not be following the site rules.

However, a more realistic approach needs to be taken as turning the contractor away may affect the programme.

A full review with the contractor is necessary, i.e.:

- What needs to be done?
- Who needs to do it?
- Where does it need to be done?
- Where are the Risk Assessments?
- Where are the Method Statements?
- What work may affect others in the area?
- What controls for safety does the contractor propose using (e.g. RCDs)?
- How does he propose to manage the hazard of 230 v tools and cables?
- What Permit to Work methods does he propose?
- Can the work area be isolated from others so as not to expose others to unnecessary hazards?

If the Risk Assessments and Method Statements demonstrate that a safe system of work can be operated, that the risks to *all* operatives can be controlled, that the work is of relatively short duration then it *may* be possible to allow the work to go ahead.

Using 230 v on a construction site is a serious hazard which poses great risk of electrocution, electric shock, electric burns and fire, but it is *not* against the law and provided the risks can be managed it may be permissible.

However, best practice and all safety guides advocate the use of 110 v power tools or cordless 12 v power tools.

Does all electrical equipment have to have an inspection and test certificate?

The Electricity at Work Regulations 1989 requires the 'duty holder' (usually the employer) to ensure that electrical equipment is:

'maintained in a safe condition'.

The Regulations do not specifically require elaborate testing and inspection, but if these steps are needed to ensure that the equipment is safe then they must be carried out.

A simple, and usually effective, procedure to check electrical equipment for faults and damage is visual inspection. Approximately 95% of faults or damage can be detected by carrying out a thorough visual inspection.

All operatives should be trained on what to look for when carrying out a visual inspection and everyone should be encouraged to use the techniques *every* time they use electrical equipment.

Before an electrical hand tool, RCD, lead or cable extension, etc. is used, check that:

- no bare wires are visible
- the cable covering is not damaged
- the cable cover is free from cuts and abrasions
- the plug is in good condition
- the plug pins are not bent or missing, the casing cracked, etc.
- there are no joined cables, with insulating or other tape — any cable connectors must be proprietary
- the outer cable covering (sheath) is gripped properly where it enters the plug or equipment (e.g. no coloured wires showing)
- the equipment itself is in good condition with no parts missing, screws missing, casing cracked, etc.
- there are no overheating marks, burn or scorch marks around the plug, socket or equipment
- the RCD is working correctly.

Some equipment will require more definitive evidence that it has been formally inspected and a certificate is a good record.

Equipment which has a high risk of becoming unsafe through misuse should be inspected and tested, usually using a PAT test. This test will check the earth continuity.

If equipment is going to be used in hazardous areas (e.g. confined spaces), it should be inspected and tested regularly, and suitable records kept.

If equipment is tapped to earth it should be regularly checked to ensure that the earth wire is intact and connected. Abuse of 110 v tools by multi-users will often cause wiring to come loose. This may not be seen by visual inspection. A formal process of combined testing would be sensible.

Top tips

Equipment	Voltage	User checks	Formal visual inspection	Combined inspection and testing
Battery-operated power tools	Less than 25 v, usually 12 v	No	No	No
Torches	Less than 12 v	No	No	No
25 v portable hand lamps	25 v, powered from transformer	No	No	No
50 v portable hand lamps	50 v, secondary winding, centre tapped to earth	No	No	Yearly

Table continued overleaf

Equipment	Voltage	User checks	Formal visual inspection	Combined inspection and testing
110 v portable hand tools, extension leads, site lighting	Max. 55 v tapped to earth	Weekly	Monthly	Before first use on site, then three-monthly
230 v portable and hand-held tools	230 v mains supply	Daily/every shift	Weekly	Before use and then monthly
230 v portable floodlighting, extension leads	230 v	Daily	Weekly	Before first use, monthly
230 v 'fixed' equipment, e.g. lifts, hoists, permanent lighting	230 v	Weekly	Monthly	Before first use and then three-monthly
RCDs portable		Daily, every shift	Weekly	Before first use and then monthly
RCDs fixed		Daily	Weekly	Before first use and then three-monthly
Mains equipment in site offices, e.g. fax machines, photocopiers	230 v	Monthly	Six-monthly	Before first use, then yearly

Source: HSE: HSG 150.

Are there any additional precautions which need to be taken when using electrical equipment in hazardous areas?

Hazardous areas will include:

- confined spaces
- flammable atmospheres (e.g. paint shops)
- chemical plants
- areas subject to dust and particle accumulation.

Before any work is carried out in these areas a Permit to Work should be issued. This will detail all of the safety precautions to be taken for working in the area, including which type of electrical power tool is appropriate.

In any hazardous environment, 12 v portable tools are safer to use than any other.

What additional steps need to be taken when working near overhead power lines?

Many deaths and serious injuries are caused when operatives come into contact with overhead power lines.

Work must be carefully planned to avoid accidental contact.

Wherever possible, the electrical power supply to the overhead cables should be disconnected and isolated.

Often, contact with overhead power lines is inadvertent because the operative is unaware of the risks — often the assumption is made that the lines are dead.

Carrying long scaffold tubes, handling long metal roof sheets, carrying long ladders, using lifting plant, mobile elevating work platforms — all can contribute to the likelihood of hitting an overhead power line with fatal consequences.

A Permit to Work system should operate where overhead power lines exist.

Work should, where possible, be away from the lines.

The lines may be temporarily switched off.

Plan what works need to be carried out in the area of overhead power cables — measure the height of the cables and specify that any long pieces of material or equipment shall be less than the height of the cables, including a safe tolerance area.

Prevent vehicle access to overhead power lines by creating barriers and restricted areas.

Are there any special precautions to take in respect of health and safety when commissioning and testing of fixed plant and equipment takes place prior to project completion?

Equipment will need to be commissioned during the course of the development works and this is likely to involve the use of 230 v mains power supply.

There is a health and safety risk when using a 230 v supply on a construction site and the following procedures should be adopted.

- The Principal Contractor should be required to produce an agreed plant commissioning programme which will incorporate information about the site and from other sub-contractors/nominated contractors who require plant and equipment to be commissioned.
- All sub-contractors will be informed about the commissioning of plant and equipment and the Principal Contractor shall display a series of notices on either the individual equipment or on the entrance to the area which advises everyone that the equipment is now live and connected to 230 v.

- Secure means of isolation shall exist for each part of the installation on which work is carried out. Padlocks and keys should be clearly identified and held by the Principal Contractor.
- Circuits which are not in use, i.e. used for commissioning and thereafter not required until the site is handed over, should be locked off and doors to switch rooms, fuseboxes, etc. should be locked shut with keys kept by the Site Agent.
- All electricians should be suitably trained to work with high voltages and, obviously, the Electricity at Works Regulations 1989 must be complied with.
- Each employer will be responsible for his own employees and must ensure that suitable information, instruction and training has been provided, that their employees are competent to do the job, etc.
- The Principal Contractor will be responsible for ensuring that site safety rules are followed and that all contractors and sub-contractors are kept informed of all site safety matters.

Mechanical plant and equipment (e.g. air handling units, lifts and escalators) will be operating from mains voltage while the site is still a construction site. The hazards and risks of 230/240 v on site must therefore be properly and safely managed.

All cables, ductwork, etc. carrying 230/240 v power must be labelled with a hazard warning sign (yellow and black) as follows:

'Warning: Live 230/240 v power
Do not touch'

or similar.

All electrical cables running adjacent to work areas *shall* be encased in armour-plated conduit (or other recognised material) and sufficiently and visibly labelled:

'Live electrics'

The Principal Contractor shall ensure that *all* operatives on site are aware that the mechanical and electrical equipment may be operational. Tool-box talks and notices displayed in mess rooms will provide timely reminders.

References

Electricity at Work Regulations 1989.

Health and Safety at Work Etc. Act 1974.

Provision and Use of Work Equipment Regulations 1998.

Electrical Safety on Construction Sites: HSG 141.

Maintaining Portable and Transportable Electric Equipment: HSG 107.

Avoidance of Danger from Overhead Power Lines: Guidance Note GS6.

Requirements for Electrical Installations: BS 7671.

11

Hazardous substances

What are the legal requirements governing the use and exposure to hazardous substances?

There are a number of legal requirements governing the use and exposure to substances, liquids, vapours, gases, fumes and biological agents which may be hazardous to health.

Hazardous to health means having the potential to cause harm to the health of a person.

Harm to health can be by:

- ingestion
- inhalation
- absorption through the skin.

The main pieces of legislation are:

- Control of Substances Hazardous to Health Regulations 2002 (COSHH)
- Control of Asbestos at Work Regulations 2002
- Control of Lead at Work Regulations 2002.

Generally, an employer must not expose employees or others to substances hazardous to health unless a Risk Assessment has been carried out — usually known as a COSHH Assessment.

Good practice in respect of managing exposure to hazardous substances makes reference to 'the hierarchy of risk control', namely:

- eliminate the use of hazardous substances
- substitute for a lesser risk substance
- control substance at source
- provide personal protective equipment
- monitor and review controls.

Employers must remember that when two or more 'harmless' substances are mixed together they may combine to form a hazardous substance.

What are some of the health effects of hazardous substances?

Examples of the effects of hazardous substances include:

- skin irritation and dermatitis
- asthma
- respiratory conditions
- loss of consciousness as a result of being overcome by toxic fumes
- cancer
- infection from bacteria.

What is a hazardous substance under COSHH?

If a substance or mixture of substances is classified as dangerous under the Chemicals (Hazard, Information and Packaging for Supply) Regulations (CHIP) then the COSHH Regulations apply to it.

Dangerous substances can be identified by their warning label and the supplier must provide safety data sheets for all hazardous substances.

The Health and Safety Executive publish lists of the most commonly used dangerous substances.

Substances which have an Occupational Exposure Limit are classified as hazardous. These are substances which can be considered 'safe' up to certain exposure limits.

Any kind of dust in concentrations specified in the Regulations is deemed to be harmful.

COSHH does *not* cover asbestos and lead as these are dealt with under specific legislation.

What types of substances in use on a construction site could be considered as hazardous?

The following are commonly in use on construction sites and could be considered as a hazardous substance under COSHH:

- solvents
- silica
- cement
- mineral oils
- carbon monoxide gas
- carbon dioxide gas
- adhesives
- paints
- wood dusts
- welding fumes
- carbon deposits and soot
- acid cleaners
- detergents and degreasers
- pesticides.

Case Study

Two lift engineers were working in the lift shaft, welding the metal frame. They were adequately protected by personal protective equipment and had a local exhaust ventilation system which removed the welding fumes. However, unbeknown to them, the exhaust extract was faulty because it had not been subjected to regular inspection and the welding fumes were leaking into the roof void above the lift shaft in which two operatives were laying electrical cables. Both operatives in the roof void were overcome by fumes and were found when the Site Agent checked on progress of works.

Details from the suppliers or manufacturers must be obtained for all substances planned to be used on site, or brought in by others to be used on site.

Remember, while the operatives in the immediate vicinity of the hazardous substance may be protected by personal protective equipments, others working some distance away could be affected by the substances, fumes, vapours, mists or dust.

How do hazardous substances affect health?

All types of hazardous substances will affect the body in different ways, depending on how the substance enters the body — known as the route of entry.

Inhalation

Inhalation of a hazardous substance probably accounts for the majority of deaths associated with using substances.

Inhalation of dusts, fumes, mists or vapours will affect the health of an individual in different ways and at different speeds.

Acute exposure refers to the immediate on-set of symptoms, i.e. the exposure is so great that an immediate adverse reaction happens (e.g. unconsciousness, acute respiratory failure, heart attack, etc.).

Chronic exposure is exposure to substances over a prolonged period of time where the harmful substance accumulates in the body over that period of time — frequently through 'little and often' exposure.

Lung diseases often occur because of persistent and long-term exposure to the hazard (e.g. miners and coal dust).

Ingestion

Substances, once swallowed, will enter the stomach and intestines and pass into the bloodstream. When this happens, the toxic substances are transported around the body to all major organs.

Ingestion mostly occurs from 'hand to mouth' contact, i.e. eating or drinking with contaminated hands. Also, if drops of the harmful substance or dust, etc. are accidentally transferred on to the mouth and the person wipes their mouth they could transfer the substance onto the tongue where it is swallowed, etc.

Absorption through the skin

Certain substances and micro-organisms can pass through the skin which acts as a protective barrier to underlying tissue and the bloodstream.

Substances can therefore enter into tissue and the bloodstream and then travel around the body to vital organs. Some substances are so hazardous that only minute droplets are needed to come into contact with the skin (e.g. Ricin — a common terrorist chemical weapon) in order to cause multiple organ failure and death.

Any cuts or grazes to the skin, abrasions, etc. will increase the risk of absorption of a chemical through the skin.

Sometimes the skin does not allow complete transference of the substance into underlying tissue and the bloodstream and an allergic reaction takes place. This manifests itself in visible skin irritation, rashes, blisters, burns, etc.

The most common parts and organs of the body affected by hazardous substances are:

- lungs
- liver
- bladder
- skin
- brain
- central nervous system
- circulatory system
- reproductive system
- urino-genital system.

The most common illnesses suffered by construction workers are:

- lung diseases (e.g. pneumoconiosis)
- lung cancer
- silicosis
- dermatitis
- asthma
- leptospirosis
- legionnaires' disease.

What are the steps which need to be followed in order to produce a COSHH Assessment?

Regulation 6 of the Control of Substances Hazardous to Health Regulations 2002 requires employers to carry out a 'suitable and sufficient' assessment of the risks to health from using the hazardous substance.

'Suitable and sufficient' does not mean absolutely perfect, but the guidance which supports Regulation 6 lists a number of considerations which must be taken into account during the process.

The first step in providing a suitable and sufficient COSHH Assessment is to ensure that the person carrying it out is *competent*.

Regulation 12(4) of COSHH requires any person who carries out duties on behalf of the employer to have suitable information, instruction and training.

A competent person does not necessarily need to have qualifications as such but they should:

- have adequate knowledge, training, information and expertise in undertaking the terms 'hazard' and 'risk'
- know how the work activity uses or produces substances hazardous to health
- have the ability and authority to collate all the necessary relevant information
- have the knowledge, skills and experience to make the right decisions about risks and the precautions that are needed.

The person carrying out the assessment does not need to have detailed knowledge of the COSHH Regulations but needs to know who and where to go to in order to obtain more information. They need to be able to recognise when they need more information and expertise. As the saying goes 'a little knowledge is a dangerous thing'.

The COSHH Assessment process can be broken down into seven steps as follows.

Step 1: Assess the risks

Identify the hazardous substances present on the site, or intended to be used on the site.

Consider the risks these substances present to your own employees and all others.

List down all of the substances likely to be used. If the substances are on site, read the labels and look for the hazardous warning symbols as follows:

Obtain safety data sheets from the supplier or manufacturer.

A safety data sheet must be provided by the manufacturer or supplier and it must give information on, among other things, common usage, constituent ingredients, exposure limits, personal protective equipment, emergency procedures, first aid, spillage precautions.

If sub-contractors are supplying the hazardous substances then they must be required to submit manufacturers' data sheets together with their COSHH Assessments.

The COSHH Regulations 2002 introduced a revision to the information needed for a COSHH Assessment and now *all* COSHH Assessments must have attached to them the product's safety data sheet.

Consider the risks of the hazardous substances identified to people's health. Remember, it is all people and not just employees. Again, the 2002 COSHH Regulations clearly state that the effect on health of a hazardous substance to 'other persons' must be clearly considered.

The assessing of risk to health from using the hazardous substance is a matter of judgement. Sometimes it is not possible to know for sure what level of exposure will be harmful. In these cases, it is preferable to always err on the side of caution.

Some questions to consider are:

- How much of the substance is in use?
- How could people be exposed?
- Who could be exposed to the substance and how often?

- What type of exposure will they have?
- Could other people be exposed to the harmful substance?

Step 2: Decide what precautions are needed

The first responsibility for an employer is to eliminate the risk of using a harmful substance. This is an effective precaution to take but may not always be possible.

Next, consider whether an alternative substance could be used which is less hazardous than the original substance. If there is something on the market which does the job more safely then it should be used.

If the substance cannot be eliminated or substituted, then it must be used with *suitable controls* so as to reduce any level of exposure to an acceptable limit.

Suitable controls may be:

- changing the way the work is done (e.g. painting instead of spraying)
- reducing concentrations of the substance
- modifying the work process (e.g. to reduce exposure time)
- reducing the number of employees and others exposed to the substance
- adopting maintenance controls and procedures
- reducing the quantities of substances kept on site
- controlling the overall working environment (e.g. increasing ventilation, damping down dust, etc.)
- adopting appropriate hygiene measures (e.g. easily accessible hand/arm washing stations to prevent skin absorption)
- enclosing the work activity.

Step 3: Prevent or adequately control exposure

Having decided what controls are required, the next step is to implement the controls on the site.

The very nature of a construction site makes it quite difficult to implement some controls but ones to consider will be:

- substitution of substance
- exclusion of workers not needed from the area
- introduction of local exhaust ventilation
- changing work processes (e.g. reduce spraying in favour of brush application)
- creating zoned work areas
- undertaking tasks where practicable in the open air.

Construction sites are particularly prone to a 'cocktail of substances' whereby different substances which may combine to form a separate harmful substance. The Principal Contractor should co-ordinate the use of these different substances and should complete the COSHH Assessment for any 'communally' produced hazardous substance.

If there are no effective alternatives, it is permissible to issue operatives with personal protective equipment.

Personal protective equipment should be considered as the last resort — all other control measures must have been effectively considered and introduced if at all possible.

Personal protective equipment will include:

- face masks
- respiratory masks
- gloves or gauntlets
- goggles
- safety boots or shoes which are chemical resistant.

Control of harmful substances *must* be to a level that most workers can be exposed to day after day without adverse effects on their health.

Adequate control of harmful substances can be defined by referring to the Occupational Exposure Limit of a substance (OEL).

An OEL is set at a limit that is not likely to damage the health of workers who are exposed to it day after day.

If a substances has an OEL and the level of exposure is kept to the limit stated then, as an employer, you will be deemed to have adequately controlled the risk.

Short-term, infrequent exposure to higher levels than the OEL are permissible but they must be exceptional rather than the norm. Usually, such higher exposure will be because of an emergency, e.g. spillage, substance release, failure of exhaust ventilation, etc.

Maximum Exposure Limits (MEL) are set for substances which may cause the most serious health effects such as cancer or occupational asthma.

Any substance with an MEL must be reduced to a level which is below the MEL, and employers are required to reduce the level of exposure to the substances to as *low as is reasonably practicable*.

Step 4: Ensure that control measures are used and maintained

Employees and others must make proper use of the controls which an employer puts into place.

The Principal Contractor on a multi-occupied site must ensure that contractors follow their own COSHH Assessments and that others affected by the works also adhere to the controls identified as necessary.

An employer has a duty to provide personal protective equipment (PPE) to employees and to ensure that it is suitable and sufficient for their needs and is properly maintained.

Employees must report defects to PPE to the employer, or, if PPE has been provided by the Principal Contractor, to him or his representative.

All controls introduced onto the site to reduce the exposure to harmful substances must be adequately maintained so as to ensure that they are effective.

Engineering controls and local exhaust ventilation equipment must be regularly inspected, tested and examined and records shall be kept. Local exhaust ventilation must be inspected every 14 months.

Records and test certificates must be kept for at least five years. The Principal Contractor should keep these for any communal controls introduced into the site.

Step 5: Monitor exposure

If the Risk Assessment determines that there is a serious risk to health if people are exposed to a substance then health surveillance must be considered.

If any of the following apply, health surveillance will be essential:

- if control measures fail or deteriorate
- if exposure limits could be exceeded
- if control measures are not maintained adequately.

Air monitoring must be carried out if there is a risk of exposure to the harmful substance due to inhalation.

Air monitoring is either site-wide with environmental air monitoring or can be achieved by giving exposed workers personal air monitors.

Records of air monitoring should be kept for at least five years.

Step 6: Carry out health surveillance

Health surveillance is defined as:

an assessment of the state of health of an employee, as related to exposure to substances hazardous to health, and includes biological monitoring.

Health surveillance will be necessary in the following circumstances:

- any exposure to lead fumes
- exposure to substances which cause industrial dermatitis
- exposure to substances which may cause asthma
- exposure to substances of recognised systemic toxicity, i.e. substances that can be breathed in, absorbed through the skin or swallowed and which affect parts of the body other than where they enter.

Any health surveillance carried out should be recorded and records should be kept for at least 40 years. It is for the *employer* to decide whether health surveillance is necessary.

Step 7: Information, instruction and training

All employees expected to use or to come into contact with substances hazardous to health shall receive suitable information, instruction and training.

The Site Agent should ensure that all contractors have adequate records for training their operatives in the health and safety risks of using hazardous substances.

The Site Agent should ensure that the site induction training covers the use of hazardous substances on the site and the control measures to be followed.

All operatives, visitors, contractors, etc. are entitled to see the COSHH Assessment and the attached safety data sheet.

Information may be way of safety notices, tool-box talks, site rules, etc.

Instruction generally covers one-to-one exchanges on how to do something, what to use, what PPE to wear, etc. The health effects of using the hazardous substance should be clearly discussed.

It is useful to keep records of any exchange of information on COSHH in the site records book.

What additional precautions are needed when dealing with exposure to lead?

Lead is a major health hazard and is controlled by its own Regulations:

- Control of Lead at Work Regulations 2002.

Managing the risks from lead on site can be quite complex, although the general principles of COSHH broadly apply.

Any exposure to lead, lead fumes or vapours, dusts, etc. must be subject to a 'suitable and sufficient' Risk Assessment.

Specific controls are required if employees are going to be exposed to 'significant exposure', i.e.:

- exposure exceeds half the occupational exposure limit for lead
- there is substantial risk of an employee ingesting lead
- there is a risk of the employee's skin coming into contact with lead alloys or alkyls.

If exposure is deemed to be 'significant' then an employer must:

- issue employees with protective clothing
- monitor lead in air concentrations
- place employees under medical surveillance
- use Risk Assessments to determine control measures
- identify any other measures to reduce exposure and comply with the Regulations.

References

Control of Substances Hazardous to Health Regulations 2002.

Control of Lead at Work Regulations 2002.

Control of Asbestos at Work Regulations 2002.

COSHH Regulations 2002: Approved Code of Practice and Guidance: L5 (2002) HSE.

Lead at Work Regulations 2002: Approved Code of Practice and Guidance: L132 (2002) HSE.

Step by Step Guide to COSHH Assessment: HSG 97: HSE.

12

Personal protective equipment

What are the responsibilities of the Site Agent with regard to issuing personal protective equipment (PPE)?

The duty to issue personal protective equipments rests with the employer or self-employed person and not with the Site Agent or Principal Contractor.

Where the Site Agent is the employer's on-site manager, then the employer has a duty to ensure that his *own* employees are correctly issued with PPE.

The Principal Contractor will determine the site rules for a construction site and in the rules it can stipulate the minimum requirement for PPE.

The Site Agent must ensure that his site rules are being followed — he has the duty to monitor and review safety on site.

Who determines whether a site is a 'hard hat' site or not?

Generally, the Principal Contractor will determine whether the site will be a hard hat site or not.

The applicable legislation is the Construction (Head Protection) Regulations 1989 and in particular Regulation 5 stipulates that the person having control of the site may make rules to ensure that others comply with the requirements of Regulation 4, which stipulates that an employer must ensure that suitable head protection is worn.

Rules about where and when head protection should be worn and by whom must be in writing and clearly displayed on the site.

Usually, all entry points, mess rooms and canteens display safety signs stating that 'hard hats must be worn'.

Where there is 'no foreseeable risk of head injury, other than by falling' hard hats do *not* need to be worn. In effect, the employer or Site Agent needs to complete a Risk Assessment which identifies the hazards and risks of working on the site.

If a Site Agent relaxes the hard hat rule, there should be a Risk Assessment which identifies that the risk of head injury is minimal.

It may be possible to relax the rule for hard hats for parts of the site, although this is not recommended as operatives may be confused about where and when to wear their hard hat.

Many fatalities and severe injuries are caused because employees and others do not wear protective hats. Brain damage is often irreversible and, although someone may be physically very able, the damage done to the brain via a head injury could leave the individual with a mentality of a child.

As a sub-contractor on site, can I decide that the area in which my employees are working does not have to be a hard hat area?

As an employer in control of your own site, you can decide when hard hats need to be worn but as a sub-contractor on a multi-occupied site you will have to follow the site rules and wear head protection.

Under the Construction (Design and Management) Regulations 1994, the Principal Contractor must make site rules with regard to the implementation and management of site safety. As a contractor, you have a duty to co-operate with the Principal Contractor and to follow the site rules.

There would be no harm in discussing the situation with the Site Agent with a view to seeking authority to relax the rules for head protection in your work area.

There may, of course, be other works being undertaken which you are not aware of and these could pose head injury hazards to your employees. It is the Principal Contractor's duty to co-ordinate and manage the hazards and risks of other contractors on a multi-occupied site.

What are the legal requirements with regard to personal protective equipment?

The Personal Protective Equipment at Work Regulations 1992 apply in all work environments.

Employers have a duty to provide to the employees, *without charge*, all necessary personal protective equipment which has been identified as necessary within the Risk Assessment.

PPE must be suitable and sufficient for use and appropriate for the risks it has been chosen to reduce. It must suit the worker for whom it is intended and afford adequate protection.

PPE must be comfortable to wear and fit properly to the user. If more than one type of PPE is to be used, they must be compatible, e.g. goggles and face mask must be suitable to be worn together.

Employers must ensure that employees have been given adequate information, instruction and training about how to wear PPE, why it has to be worn, its benefits, how to check it for defects, maintenance procedures, etc.

Suitable accommodation must be made available for the storage of PPE if it is impracticable for employees (and others) to store it on site.

Employers could be prosecuted for failing to supply appropriate PPE or for failing to maintain it in a suitable condition for use.

Is the Site Agent responsible for issuing PPE to site visitors?

It depends on the visitors! Usually, a Site Agent will provide PPE for the Client and Client's representatives as it would be unreasonable to expect the Client to provide the appropriate equipment. Hard hats are absolutely essential if there is a risk of falling objects or of hitting one's head on scaffold tubes, ceiling projections, etc.

It may not be reasonable for the Site Agent to provide PPE for:

- architects
- designers
- quantity surveyors
- building services consultants
- project managers
- planning supervisors

unless it is *specialist* PPE which is site-specific. All of the above will be employees of professional service firms or will be self-employed. The duty to provide PPE falls to the *employer*. They should provide their own!

What PPE would usually be expected on a main construction site?

The usual PPE comprises:

- hard hats
- safety boots or shoes
- hi-visibility vests and jackets
- gloves
- goggles or eye protectors.

It may be necessary to provide overalls and protective clothing as well. It may also be necessary to provide hearing protection.

Should a Risk Assessment be completed in order to determine what PPE is required?

All work activities where there is a risk of injury to employees or others should be subjected to a full Risk Assessment.

Under the Management of Health and Safety at Work Regulations 1999, employers are responsible for completing Risk Assessments and for recording their significant findings.

Employers should therefore conduct Risk Assessments for any construction site based activity and issue appropriate PPE.

The Site Agent should consider the hazards and risks in communal areas on the site and set down the findings in the Risk Assessment. From this, the site rules will be drawn up which stipulate what PPE is to be worn, when and by whom.

The Site Agent may need to issue PPE to persons who are exposed to risk from hazardous substances from the activities of others.

References

Personal Protective Equipment at Work Regulations 1992.

Personal Protective Equipment at Work Regulations 1992 — Approved Code of Practice.

A Short Guide to the Personal Protective Equipment at Work Regulations 1992: INDG 174: HSE.

Personal Protective Equipment: Safety Helmets: CIS 50: HSE.

13

Manual handling

What are the main hazards and risks associated with manual handling activities?

Manual handling activities are responsible for millions of lost time accidents and injuries and over 50% of all 'over three day' accidents involve manual handling of some sort.

'Bad backs' are common and are not always caused by lifting exceptionally heavy weights. Often it is an accumulation of inappropriate lifting techniques, bad posture and poor systems of work. Sometimes, bending down to pick up a pencil will be sufficient to cause a muscular spasm, slipped disc or other injury.

The main hazards are:

- lifting too heavy a weight
- lifting awkward weights
- tripping while carrying
- pushing, pulling and shoving loads
- repetitive lifting
- repetitive actions
- dropping weights or loads.

The hazards, i.e. the activities with the potential to cause harm, will often result in the following injuries:

- slipped discs
- pulled muscles
- strained muscles
- torn muscles
- torn ligaments
- cracked bones (e.g. ribs)
- repetitive strain injuries
- impact injuries
- hernias.

What are the legal requirements regarding manual handling activities?

The Manual Handling Operations Regulations 1992 set down the legal duties surrounding manual handling activities at work.

Employers are required to complete Risk Assessments for all manual handling activities carried on at work, and the significant findings must be recorded in writing.

The responsibility under the Regulations for the employer is to eliminate or reduce the amount of manual handling undertaken by employees.

The risk of injury must be reduced to the lowest level possible.

What are some of the practical steps which can be taken to reduce manual handling on construction sites?

The first duty to reduce manual handling activities on construction sites sits with Designers under the Construction (Design and Management) Regulations 1994.

Designers have a duty to design out hazards associated with their designs both in relation to the future use of the building and to how it will be constructed.

The Planning Supervisor should assist in ensuring that the Designers have considered their duties and provided Design Risk Assessments for the Principal Contractor.
Designers should consider:

- the weight of materials they specify
- the size of the materials they specify
- the type of materials they specify
- the sequence of construction.

The Planning Supervisor should encourage Designers to reduce the likelihood of injury from manual handling activities. Consideration should also be given to how materials will be delivered and moved around site, e.g. carrying 3 m × 2 m boards up ten flights of stairs may be excessive manual handling and the Designer should have considered smaller boards or hoist access into the building.

Materials should be available in the lightest weights possible. Bags, blocks, etc. should weight 25 kg or less.

Once the Principal Contractor is in receipt of any Design Risk Assessments and the Pre-tender Health and Safety Plan (as this may outline the philosophy of the design in respect of manual handling), the Site Agent should carry out a Risk Assessment of the materials needed on site, the sizes specified in the design, access and movement of materials around the site and to the locations needed.

Manual handling must be eliminated where possible and in any event reduced to acceptable levels.

Mechanical aids, lifts, hoists, trolleys, etc. eliminate most of the manual handling activity and the Risk Assessment should consider these options.

Plan what equipment is needed throughout the site.

Avoid double handling of materials — it increases the risk of injury.

Reduce the height to which materials need to be lifted by positioning loads at height by mechanical aids.

Reduce the distance materials need to be carried.

Eliminate the 'macho' culture which reinforces the message among operatives that it is 'weak' to ask for assistance.

Encourage the sharing of lifting of loads which are too heavy for one person or which are of an awkward shape. Two people lifts will usually be safer than one person lifts.

How is a manual handling Risk Assessment carried out?

There are four key elements to undertaking a manual handling Risk Assessment, namely:

- consider the tasks
- consider the loads
- consider the working environment
- consider the individual capability.

The tasks

Do they include:

- holding or manipulating loads at a distance from the trunk
- unsatisfactory bodily movement or posture
- twisting the trunk
- stooping
- reaching upwards
- excessive movement of loads, especially the following:
  - excessive lifting or lowering distances
  - excessive carrying distances
- excessive pushing or pulling of loads
- risk of sudden movement of loads
- frequent or prolonged physical effort
- insufficient rest or recovery periods
- a rate of work imposed by a process.

The loads

Are they:

- heavy
- bulky or unwieldy
- difficult to grasp
- unstable or with contents likely to shift
- containing sharp edges, protruding nails, screws, etc.
- hot from process.

The working environment

Are there:

- space constraints due to small work areas, restricted heights, etc. which will prevent good lifting postures
- uneven, slippery, defective, unstable floors
- changes in floor level either via steps or ramps
- extreme temperature changes which could affect physical exertion
- ventilation or air changes which could raise dust, cause gusts of wind, etc.
- overcrowding of work spaces
- physical obstructions
- poor lighting.

Individual capacity

Does the job:

- require unusual strength
- require special knowledge of the load
- create any other hazards to the individual carrying the load.

The information collected in the stages above will form the basis of assessing the hazards and risks from the manual handling task to be undertaken.

The likelihood of injury will range from high to low depending on the inter-relationship of the above factors.

Control measures will be identified to reduce the risk of injury, e.g.:

- remove obstructions
- increase lighting
- reduce load weight
- use mechanical aid
- provide training on lifting techniques.

When the control measures are implemented, the risk of injury will reduce.

The Risk Assessment will need to be reviewed regularly for the same manual handling process.

What is the correct manual handling technique?

There are several key rules for safe manual handling. Information should be freely available on safe lifting techniques in the mess room, site office, etc.

The fundamentals of safe lifting are:

- take a secure grip
- use the proper feet position — feet apart with the leading foot pointing in the direction you intend to go
- adopt a position with bent knees but a *straight* back
- keep arms close to the body
- keep head and chin tucked in
- keep the load close to the body
- use body weight where possible
- push up for the lift using the thigh muscles.

Operatives should also consider the environment they are to lift in and pre-plan any interim rest positions, how to change direction, etc.

What topics should be covered in a tool-box talk?

An effective tool-box talk agenda will cover:

- what checks to carry out before starting manual handling
- how to judge your capability
- environmental conditions
- wearing of personal protective equipment, e.g. gloves and safety boots
- carrying out a trial lift first
- getting help — it is not seen as being a wimp
- good handling techniques
- reducing the weights of objects
- using mechanical aids
- checks to carry out during the manual handling tasks, e.g. obstructions in the travel route, etc.

References

Manual Handling Operations Regulations 1992.

Manual Handling: Solutions You Can Handle: HSG 115.

Backs for the Future: Safe Manual Handling in Construction: HSG 149: HSE.

Getting to Grips with Manual Handling: A Short Guide for Employers: INDG 143 (revised).

Manual Handling Operations Regulations 1992: Guidance on Regulations: L23.

Handling Heavy Building Blocks: CIS 37: HSE.

14

Fire safety

What fire safety legislation applies to construction sites?

The following legislation may apply in some form or another to fire safety on construction sites:

- Fire Precautions Act 1971
- Fire Precautions (Workplace) Regulations 1997 and 1999
- Construction (Design and Management) Regulations 1994
- Construction (Health, Safety and Welfare) Regulations 1996
- Fire Certificates (Special Premises) Regulations 1976
- Management of Health and Safety at Work Regulations 1999
- Dangerous Substances and Explosive Atmospheres Regulations 2002.

Construction sites are places of work and therefore the duties placed on employers to maintain their premises in a safe condition apply to sites.

Construction sites carry a high risk of fire as the nature of the work increases hazards and risks which contribute to a fire starting.

What causes a fire to start?

A fire needs the following to start:

- fuel
- ignition
- oxygen.

The above three component parts are often referred to as the 'Triangle of Fire'.

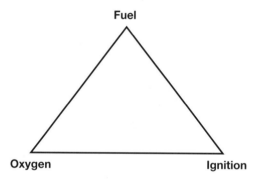

Fuel

Oxygen   **Ignition**

If one or more of the component parts of a fire are eliminated, a fire will not start or continue.

As the elements of fire are not successfully managed on all construction sites, the industry suffers approximately 4000 fires a year, costing millions of pounds in destroyed buildings, materials, plant and equipment and delayed projects.

Other insurance cover for buildings and construction work is difficult to obtain without a proper fire safety management plan in place.

As a Site Agent, what do I need to be able to demonstrate in respect of site fire safety management?

Generally, you need to demonstrate to HSE Inspectors and insurance assessors the following:

- recognition of the fire risks in the workplace and the extent of the risks
- assessment of the controls necessary to reduce the risks
- implementation of the controls necessary
- importance of constant review and monitoring of processes, controls, etc.

What is a Fire Safety Plan?

The Fire Safety Plan should identify fire risks throughout the site, e.g.:

- combustible materials
- use of hot flame equipment
- use of liquid petroleum gas
- use of combustible substances
- storage and use of any explosive materials and substances
- sources of ignition (e.g. smoking)
- use of heaters.

Once the potential fire risks are identified, i.e. where, when, why and how a fire *could* start on site (or the surrounding area, yards, outbuildings), the Fire Safety Plan should include precautions and procedures to be adopted to *reduce* the risks of fire. These could include:

- operating a Hot Works Permit system
- banning smoking on site in all areas other than the approved mess room
- controlling and authorising the use of combustible materials and substances
- providing non-combustible storage boxes for chemicals
- minimising the use of liquid petroleum gas and designating external storage areas

- controlling the siting and use of heaters and drying equipment
- operating a Permit to Work system for gas and electrical works.

Having identified the potential risks and the ways to minimise them, there will always be some residual risk of fire. The Fire Safety Plan should then contain the Emergency Procedures for dealing with an outbreak of fire, namely:

- types and locations of Fire Notices
- the location, number and type of fire extinguishers provided throughout the site
- the means of raising the alarm
- identification of fire exit routes from the site and surrounding areas
- access routes for emergency services
- procedure for raising the alarm
- assembly point or muster point.

The Fire Safety Plan should also contain the procedures to be put in place on site to protect against arson, e.g.:

- erection of high fencing or hoarding to prevent unauthorised entry
- fenced or caged storage areas for all materials, particularly those which are combustible
- site lighting (e.g. passive infra-red, security sensors)
- use of CCTV
- continuous fire checks of the site, particularly at night if site security is used.

Procedures for the storage and disposal of waste need to be included as waste is one of the highest sources of fire on construction sites.

Materials used for the construction of temporary buildings should be fire protected or non-combustible whenever possible, e.g.

providing 30-minute fire protection. The siting of temporary buildings must be considered early in the site planning stage as it is best to site them at least 10 m away from the building being constructed or renovated.

Having completed the Fire Safety Plan, a sketch plan of the building indicating fire points, assembly point, fire exit routes, emergency services access route to site, etc. should be completed and attached to the Plan. The sketch plan (which could be an Architect's existing outline drawing) should be displayed at all fire points and main fire exit routes and must be included in any site rules or information handed out at induction training.

What is the most important part of fire safety on the construction site?

Fire safety needs to be managed on a construction site in the same way that hazards and risks associated with other activities need to be.

The same concept is applied to managing fire safety as to other site hazards:

- Fire Risk Assessment.

The Fire Precautions (Workplace) Regulations 1997 (amended in 1999) apply to *all* employers and all workplaces and require employers to carry out Fire Risk Assessments and to record the significant findings in writing.

What is a Fire Risk Assessment?

A Fire Risk Assessment is a structured way of looking at the hazards and risks associated with fire and the products of fire (e.g. smoke).

A hazard is the potential to cause harm. A risk is the likelihood that it will happen.

Like all Risk Assessments, a Fire Risk Assessment follows five key steps, namely:

- identify the hazards
- identify the people and the location of people at significant risk from a fire, i.e. who they are and where they work
- evaluate the risks — how severe will things be
- record findings and actions taken to reduce any fire risks
- keep the Assessment under review.

So, a Fire Risk Assessment is a record that shows you have assessed the likelihood of a fire occurring in your workplace, identified who could be harmed and how, decided on what steps you need to take to reduce the likelihood of a fire (and therefore its harmful consequences) occurring. You have recorded all these findings regarding your construction site into a particular format, called a Risk Assessment.

How do I do a Fire Risk Assessment and how often?

A fire needs three things to start:

- oxygen
- fuel
- heat or ignition source.

If any one of these is missing, a fire cannot start.

Once a fire has started it spreads and engulfs other sources of fuel. Also, as a fire intensifies it gives off more heat. In turn, this heat causes things to ignite.

Step 1: Identify the hazards

Sources of ignition

You can identify the sources of ignition on your site by looking for possible sources of heat that could get hot enough to ignite the material in the vicinity.

Such sources of heat/ignition could be:

- smokers' materials
- naked flames (e.g. fires, blow lamps, etc.)
- electrical, gas or oil-fired heaters
- Hot Work processes (e.g. welding, asphalting, use of LPG)
- cooking, especially frying in the mess room
- faulty or misused electrical appliances including plugs and extension leads
- lighting equipment, especially halogen lamps
- hot surfaces and obstructions of ventilation grills causing heat build-up
- poorly maintained equipment that causes friction or sparks
- static electricity
- arson.

Look out for any evidence that things or items have suffered scorching or overheating, e.g. burn marks, cigarette burns, scorch marks, etc.

Check each area of the site systematically:

- access routes
- work areas
- storage areas
- messing facilities
- welfare facilities
- fuel storage areas.

Sources of fuel

Anything (generally) that burns is fuel for a fire. Fuel can also be invisible in the form of vapours, fumes, etc. given off from other less flammable materials.

Look for anything on the site that is in sufficient quantity to burn reasonably easily, or to cause a fire to spread to more easily combustible fuels.

Fuels to look out for are:

- wood, paper, cardboard
- flammable chemicals (e.g. cleaning materials)
- flammable liquids (e.g. cleaning substances, liquid petroleum gas)
- flammable liquids and solvents (e.g. white spirit, petrol, methylated spirit)
- paints, varnishes, thinners, etc.
- furniture, fixtures and fittings
- textiles
- ceiling tiles and polystyrene products
- waste materials, general rubbish
- gases.

Consider also the construction of the building if appropriate — have any materials been used which would burn more easily than other types. Hardboard, chipboard and blockboard burn more easily than plasterboard.

Identifying sources of oxygen

Oxygen is all round us in the air that we breathe. Sometimes, other sources of oxygen are present that accelerate the speed at which a fire ignites, e.g. oxygen cylinders for welding.

The more turbulent the air, the more likely the spread of fire will be, e.g. opening doors brings a 'whoosh' of air into a room and the fire is fanned and intensifies. Mechanical ventilation also moves air around in greater volumes and more quickly.

Do not forget that while ventilation systems move oxygen around at greater volumes, they will also transport smoke and toxic fumes around the building.

Step 2: Identify who could be harmed

You need to identify who will be at risk from a fire and where they will be when a fire starts. The law requires you to ensure the safety of your staff and others (e.g. contractors). Would anyone be affected by a fire in an area that is isolated? Could everyone respond to an alarm, or evacuate?

Will you have people with disabilities on the site, e.g. visually or hearing impaired? Will they be at any greater risk of being harmed by a fire than other people?

Will contractors working in plant rooms, on the roof, etc. be adversely affected by a fire? Could they be trapped or not hear alarms?

Who might be affected by smoke travelling through the building or site? Smoke often contains toxic fumes.

Step 3: Evaluate the risks arising from the hazards and the control measures you have in place

What will happen if there is a fire? Does it matter whether it is a minor or major fire?

A fire is often likely to start on a construction site because:

- people get careless with cigarettes and matches in case they get 'caught'
- people purposely set light to things
- Hot Works are not controlled
- people put combustible material near flames or ignition sources
- equipment is faulty because it is not maintained (e.g. electrical tools)

- poor electrical safety is allowed to be the norm in site offices and mess rooms.

Will people die in a fire from:

- flames
- heat
- smoke
- toxic fumes?

Will people get trapped in the building?

Will people know that there is a fire and will they be able to get out?

Step 3 of the Risk Assessment is about looking at what *control measures* you have in place to help control the risk or reduce the risk of harm from a fire.

Remember — Fire Safety is about Life Safety. Get people out fast and protect their lives. Property is always replaceable.

You will need to record on your Fire Risk Assessment the fire precautions you have in place, i.e.:

- What emergency exits do you have and are they adequate and in the correct place?
- Are they easily identified, unobstructed, clear of boxes, furniture, etc.?
- Is there fire fighting equipment?
- How is the fire alarm raised?
- Where do people go when they leave the building — an assembly point?
- Are the signs for fire safety adequate?
- Who will check the building and take charge of an incident, i.e. do you have someone appointed to manage things?
- Are fire doors kept closed but unlocked?
- Are ignition sources controlled and fuel sources managed?
- Do you have procedures to manage contractors on the site, especially those using Hot Works?

Are operatives trained in what to do in an emergency? Can they use fire extinguishers? Do you have fire drills? Is equipment serviced and checked, e.g. emergency lights, fire alarm bells, etc.

Step 4: Record findings and actions taken

Complete the Fire Risk Assessment form and keep it safe.

Make sure that you share the information with all site operatives.

If contractors come to site, make sure that you discuss *their* fire safety plans with them and that you tell them what your fire precaution procedures are.

Step 5: Keep Assessment under review

A Fire Risk Assessment needs to be reviewed regularly — about every week and whenever something has changed.

Top tips

- Complete Fire Risk Assessments.
- Check fire evacuation procedures.
- Check all fire precautions identified as being necessary are in place and working.
- Ensure operatives are trained. Share information with them. Fire safety is *everyone's* responsibility.
- Report any maintenance defects or carry out repairs — in a fire, in thick smoke and without lights, you do not want to trip over a break a leg.
- Never take risks with fire — discretion is the better part of valour!
- Vet contractors to make sure that they will work safely.
- Remember to consider arson.

What are general fire precautions on a construction site?

The term 'general fire precautions' is used to describe the structural features and procedures needed to achieve the overall aim of fire safety, which is to ensure that:

'Everyone reaches safety if there is a fire'.

Putting a fire out is secondary to 'life safety'.

Fire Risk Assessments are about ensuring that you have considered the likelihood of a fire starting, who it would affect, how quickly and how those people would be evacuated to a place of safety.

General fire precautions cover:

- escape routes and fire exits
- fire fighting equipment
- raising the alarm
- making emergency plans
- limiting the spread of fire (compartmentation).

General fire precautions will invariably differ from site to site depending on the complexity of the site.

Life safety for all persons on the site must be properly considered, planned, implemented and regularly checked. Escape routes, for instance, should be permanent, well identified, lead to a place of safety, well sign-posted and unobstructed. An ad hoc scramble down ladders or jumping off floors will not be acceptable.

What are some of the basic fire safety measures?

The Fire Safety Plan should be developed as an integral part of the Construction Phase Health and Safety Plan or overall work plan if CDM Regulations do not apply.

The very basic requirements are:

- fire exit routes from the site
- fire fighting equipment
- methods of raising the alarm
- lighting
- signage
- fire protection measures to prevent spread
- safe systems of work
- prohibiting smoking on site
- appointment of Fire Wardens
- regular checks
- emergency procedures.

Fire exit routes

Where possible, there should always be more than one exit route from a place of work. If the travel distance is more than 45 m to an exit, there must be two or more exits. This travel distance will vary depending on the risk rating of the site.

If the number of exits cannot be improved, then the risk rating of the site for fire must be reduced.

Fire exit routes must be unobstructed, clearly defined, of adequate size and width and not locked.

Doors leading onto fire exit routes should open via a push bar in the direction of travel.

No fire exit route shall lead back into the building or site.

Fire exit routes must lead to a place of safety.

Fire exit routes must, in themselves, be protected from fire by fire-protected enclosures or doors. Doors must be kept shut.

Exit routes using ladders, e.g. on scaffolding, need to be especially assessed as part of the site-specific Risk Assessment.

Fire exit routes must be clearly visible from all parts of the work area. Exit signs which meet the Health and Safety (Safety Signs and Signals) Regulations 1996 must be displayed.

If lighting is poor, use photo-luminescent signs.

All fire signage must display pictograms as a minimum. Text can also be used alongside the pictogram, as can directional arrows.

Fire fighting equipment

Suitable fire extinguishers need to be placed in appropriate locations around the site, and always at fire points near the fire exit routes.

Multi-purpose foam or powder extinguishers are suitable, but so, too, would be water and carbon dioxide. The Fire Risk Assessment should determine which type is required.

Fire fighting equipment should be visible, properly signed, inspected weekly and ready to use if needed. Operatives should not need to climb over materials, move plant, etc. to use the extinguishers.

Either a designated number of operatives in each work area or all operatives should be trained in how to use the fire fighting equipment.

Regular reassessment of the working area is needed to ensure that the locations of the fire extinguisher points are appropriate.

Methods of raising the alarm

A fully integrated alarm system would be beneficial on all sites, activated by break-glass points and linked to an alarm control panel in the site office.

However, this is not always possible and alternatives are permissible, e.g.:

- hand bells
- klaxons
- sirens
- hooters.

The alarm in use on the site should be clearly identified and all operatives *must* receive training in fire alarm procedures.

Fire alarm points must be clearly visible, easily accessible, etc.

If alarms cannot be heard in all areas of the site there must be a procedure for Fire Wardens to warn Fire Wardens on other floors, etc.

On small sites, a simple shout of 'Fire! Fire!' may be all that is needed.

If the site is multi-occupied without employers, e.g. a major department store refurbishment, then the construction site alarm system must integrate with that of other employers so that total building evacuation is occasioned as necessary.

Lighting

Emergency lighting is not necessarily required on all construction sites but if there is a risk of power failure and no natural daylight to the areas of work, emergency lighting will be essential.

A simple system of torches may suffice.

All emergency exit routes must be adequately lit at all times.

Emergency back-up lighting will activate when the main power supply fails.

Regular checks of emergency lighting will be necessary.

Signage

Signage enables people to be guided to safe places — either to emergency exit routes, safe places (e.g. refuges) or to assembly points.

Signs, where possible, should be photo-luminescent.

Signs must be visible from all work areas, non-confusing, of large enough size and accurate in the information they portray, e.g. must not lead to a dead end as a fire exit route.

Fire protection measures to prevent fire spread

Generally, it is best to try to consider floors and different areas as compartments, with fire protected, closed doors and fire protection to voids and ducts, etc.

Fire and smoke, including toxic fumes, spread rapidly. Compartmentation constrains it to one area.

Safe systems of work

Any work activity which looks as if it could increase the risk of a fire starting *must* be controlled by a Permit to Work or Hot Works Permit system.

Hot Works should be prevented whenever possible. Controls need to be localised, e.g. by provision of additional fire extinguishers, regular checks, additional Fire Wardens.

Combustible materials should be removed and flammable gases, etc. should be removed. Flashover should be considered.

Only trained operatives should carry out Hot Works or use flammable materials, etc.

Smoking on site

There is no acceptable fire safety procedure other than to ban it completely.

However, it may be permitted in mess rooms. If so, strict controls must be implemented.

Appointment of Fire Wardens

Each floor or work area should have an appointed Fire Warden, i.e. people who are trained to know what to do in the event of a fire, how to evacuate their area, raise the alarm, etc.

There should be enough Fire Wardens to cover for absences. Fire Wardens should receive regular training.

Regular checks

Daily and weekly fire safety checks are advisable on all sites. Checks are *always* necessary after Hot Works, and usually approximately

1 hour after the end of Hot Works so that any smouldering materials can be identified.

Records of fire safety checks should be kept for the duration of the project. Remember, you need to demonstrate that you know what you are doing.

Emergency procedures

These must be specific for each site and written down clearly. Emergency procedures must be displayed in prominent positions. They should include:

- type of fire alarm
- how to raise the alarm
- how to evacuate the site
- the assembly point
- the names of the Fire Wardens
- any highly hazardous areas
- storage of flammable materials
- procedures for visitors to site
- name and telephone numbers of local emergency services
- liaison with emergency services when they arrive on site.

Top tips

- Plan fire safety before works start.
- Reduce combustible materials.
- Reduce ignition sources.
- Keep fire exit routes clear.
- Display adequate and suitable fire signage.
- Train operatives in emergency procedures.
- Keep fire extinguishers on site, in suitable locations and of the correct type.

- Put in emergency lighting if possible.
- Clearly describe the fire alarm raising procedure.
- Remember — get everyone out rather than fight the fire.

References

Management of Health and Safety at Work Regulations 1999.

Fire Precautions (Workplace) Regulations 1997 and 1999.

Fire Safety in Construction Work: HSG 168.

Fire Safety: An Employer's Guide: HSE Books.

Fire Prevention on Construction Sites: Fire Protection Association: Fifth Edition: January 2000.

Appendix A

Site survey

A site survey is always necessary in order to be familiar with the hazards and risks associated with the building and the site.

A site survey will always be necessary before preparing the following:

- Pre-tender Health and Safety Plan
- Construction Phase Health and Safety Plan.

The Pre-tender Health and Safety Plan

A pro forma checklist has been produced which covers the main pieces of information that will be required for the Pre-tender Health and Safety Plan and guidance on completing this follows.

1. Describe the exact location of the premises, i.e. if the project is the refurbishment of a catering outlet within a site at a specific hospital, then state the precise location (e.g. on the ground floor of the main outpatients admissions building of St Bernards hospital, close to the main entrance and reception area).

2. Provide information on the type of building, for example, whether it is a unit within a premises (as in paragraph above), or part of a free-standing structure, etc. How many floors, approximate size, etc. and what the premises comprises.

3. State what type of project it is — refurbishment of existing premises, fit-out of newly built premises, etc.
4. What type of premises surround the site? It could be a department within the hospital, offices, shops, general public area (such as waiting area or reception in hospitals).
5. Around the site — what are the main hazards in the area? Railway lines, schools (children), overhead restrictions (power cables, etc.), watercourses, members of the public, etc.
6. Is work likely to cause a nuisance to neighbouring areas/ premises? Are there any restrictions on working hours/types of work, etc.?
7. How will deliveries get to the site? How will changes in level/floors be negotiated? If site is located within a larger building, is there a goods lift available? Is it adequate? Is there a specific area for receiving deliveries (loading bay, etc.)?
8. Can general public pass around the site? To front and rear? Could they be put at risk from site activities?
9. Describe the general condition of the building, e.g. good (as will probably be the case for most), run down, dilapidated, etc.
10. Is it below ground level and, if so, has it been tanked? Is it on a flood plain or near a river?
11. Is there any equipment/plant that needs to be removed during the project?
12. If 'no' state whether a current asbestos register is available.
13. Detail areas where fragile materials may be or are located, such as roofs, glazed areas, shopfronts, etc.
14. Is there any evidence of pests including droppings, feathers, sight of actual rodents/birds/insects, etc.
15. Are there any syringes or other equipment associated with drug abuse.
16. Where will skips be located, how operated (weight and load, permanent skip, etc.).
17. Is there a likelihood that deliveries will need to be co-ordinated with neighbouring premises. Could deliveries to neighbouring premises (either by nature or size) interfere with operations.
18. What is the nature of other works (if any)? Who is the principal contractor?
19. Applicable if floor/ground penetration forms part of scope of works.

20. Were photographs taken on site?
21. For example, fan for kitchen extract, etc.
22. Confined spaces, temporary supports, weakened structure, areas where a risk of falling over 2 m exists, etc.
23. Steel supports, will propping be required?
24. Do the proposed works require lifts/hoists for deliveries or food service. Are there existing lifts/hoists within the area?
25. Location of meters/mains stop valve, etc.
26. Sealed off and not used?

The Construction Phase Health and Safety Plan

When a Principal Contractor is appointed to a project, a site survey will be necessary to establish additional information regarding the building and site in order for the Construction Phase Health and Safety Plan to be developed.

Key information on hazards and risks regarding the project will be contained in the Pre-tender Health and Safety Plan. The Principal Contractor must develop the content of the Pre-tender Health and Safety Plan into the Construction Phase Health and Safety Plan.

A site survey should be used to gather information on the following.

- Use of the building — will the Client still have employees, customers, etc. within the building?
- What safety rules, risk assessments, etc. are available for the site or activity?
- What are the emergency procedures?
- What restrictions are there to site access?
- What will delivery routes be?
- What residual hazards are on the site/within the building?
- What site security procedures will need to be followed?
- Location of the nearest hospital, etc.
- Location of underground and above ground services.
- Use of adjacent land/buildings.
- Location of site welfare facilities.
- Access for construction workers.
- Storage areas.
- Manual handling hazards, e.g. need to handball materials.
- Location of any cranes or access equipment.

- Risk of unauthorised persons on to the site.
- Hazardous substances on the site.
- Presence of asbestos.
- Large delivery items and access routes.
- Type of heating required to the site.
- Type and location of lighting.

Much of the above information should be contained in the Pre-tender Health and Safety Plan, but, if it is not, it does not mean that the hazard, etc. does not apply to the site. The onus is on the Principal Contractor to assess site safety hazards himself and plan to avoid/control them.

Appendix B

Site set-up

To achieve a high standard of health and safety within a construction area, as well as compliance with the relevant legislation and approved codes of practice, requires the careful set-up of site. To achieve this, the following areas must be addressed and appropriate action taken.

Statutory notices

It is mandatory that the following statutory notices are displayed in a prominent position within the site to which they relate:

- notification of project on F10 where applicable
- health and safety law poster
- employer's liability insurance.

Additional signage and notices are to be displayed in a prominent location where applicable, as described below.

Site safety station

The site safety station should be located in a prominent location close to the site entrance. This contains first aid, personal protective equipment (spare) and emergency equipment for use by site operatives and visitors.

Segregation

To reduce the risk of theft, vandalism and unauthorised access to the site, the site must be secure. Adequate fencing and barriers to prevent access, but allowing emergency exit, must be in place. Signage stating 'No Unauthorised Access' should be displayed.

First aid

A large first aid kit, additional plasters and at least 1 litre of sterile eye wash should be located within the site safety station.

Signage showing the location and the name of the first aider should be displayed. It is expected that if five or more persons are present on site, a qualified first aider is present. For less than five personnel an appointed person must be present.

Personal Protective Equipment (PPE)

Although PPE should be used only as a last resort to protect employees from hazards in construction areas, it is often the most practicable solution. Hard hats, safety boots, gloves, etc. should be made available to operatives as required. The health and safety station should contain additional PPE for use by visitors and additional equipment for site crew if required. The station should also contain hard hats, ear plugs, dust masks, safety goggles and protective gloves.

Signs informing of the need for hard hats and protective footwear should be displayed.

Welfare arrangements

Sanitary accommodation should be provided as follows and must be available for use at all times. It must be kept clean, well lit and in working condition. The number of facilities and wash stations to be provided must comply with the table below:

No. of men at work	No. of water closets	No. of urinals	No. of wash stations
1 to 15	1	1	2
16 to 30	2	1	3
31 to 45	2	2	4
46 to 60	3	2	5
61 to 75	3	3	6
76 to 90	4	3	7
91 to 100	4	4	8

Fire precautions

An up-to-date copy of the fire plan must be displayed showing emergency exit routes, location of fire points (alarms and extinguishers) and procedures to be followed in the event of an emergency. Adequate fire points and emergency exit (running man) signage must be displayed and, where the site is located away from existing premises, additional signage will also be required.

It is recommended that each fire point contains at least one water or hydrospray and at least one CO_2 or dry powder extinguisher.

Emergency procedures should be explained to all operatives as part of their site induction.

Documentation

In addition to statutory displayed notices and signage already mentioned, the following documentation should be present on site:

- Construction Phase Health and Safety Plan
- accident book
- Risk Assessments/Method Statements
- a copy of the site rules.

Appendix C

Site safety notices

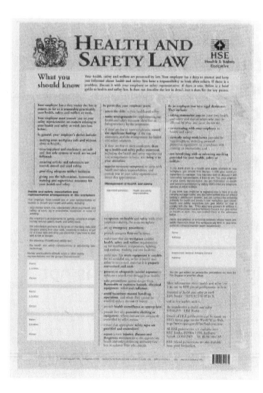

Appendix D

Health and Safety Station equipment contents

Description	Quantity	Unit
CO_2 or dry powder fire extinguisher	1	No.
Water or hydrospray fire extinguisher	1	No.
Large first aid kit	1	No.
Ear plugs (disposable)	1	Box
Face/dust masks (disposable) FFP.2	1	Box
Hard hats	3	No.
Safety goggles/glasses	2	Pair
Rubber gloves	2	Pair
Riggers gloves	4	Pair
Disposable gloves	1	Box
Waterproof plasters	1	Box
Accident book	1	No.
Eyewash (500 ml)	2	No.

Appendix E

Pre-start checklist

	Y	N	N/A	Comments
Construction Phase Health and Safety Plan issued				
Relevant method statements checked and accepted				
Relevant risk assessments (including COSHH) included in site documentation				
Clear air certificate following asbestos removal (where applicable)				
Design risk assessment completed				
HSE notification completed and sent				
All contractors approved and competent				
Site set-up agreed				
Site box fully stocked and checked				

Appendix F

Construction Phase Health and Safety Plan

Project details

Address of premises

End Client

Client

Architect

Quantity Surveyor

Structural Engineer

Building Services Consultants

Planning Supervisor

Principal Contractor

Project time-scales
- Start date:
- Completion:
- Partial handover dates:

Description of the project

Presence of asbestos on site

Site-specific conditions/restrictions that may affect safety

Maximum number of site operatives envisaged on site:

Health and Safety objectives for the project

Organisation and management for Health and Safety — on site

Site Agent

Contracts Manager

Health and Safety Manager

Safety Director

Health and Safety Consultants

Please attach specific job responsibilities for Health and Safety

Emergency procedures

Evacuation procedure

Emergency rescue procedures

Explosion, gas release

Building/structure collapse

Major chemical release

Discovery of asbestos or other prohibited substance

Scaffold collapse

Flood

Fire plan (a fire plan showing means of escape, fire points and assembly point must be appended to this Plan and a copy must be displayed on site)

Method of raising alarm (audible throughout site)

Name and number of Fire Wardens

Competent persons' names

Number and location of fire extinguishers

Means of escape

Emergency lighting

Marked-up drawing of site fire plan

Fire signage

Assembly point

Fire induction training

Specific procedures for storage of highly flammable material

Smoking policy

Provision of temporary accommodation

Fire Risk Assessments

Will LPG be used on site?   YES/NO

How and where will it be stored?

Specific control measures

What welding processes are envisaged? Include specific control measures

Site management/co-ordination

How will contractors and sub-contractors be selected and how will their competency and resources be assessed?

Who will undertake Risk Assessments on site?

Detail procedures for approving sub-contractors' Risk Assessments and method statements

How will information in relation to Risk Assessments and general health and safety issues be communicated to site operatives, sub-contractors and consultants?

What site-specific hazards and risks have been identified and how will they be controlled? You should include Risk Assessments and method statements as appropriate with your plan.

Site-specific issues

Where will the project notification (F10) be displayed?

Site-specific control measures for hazards and risks

Are any of the following activities envisaged? If so, Permit to Work systems must be implemented:

- Confined space working*
- Hot Works*
- excavations, cofferdams, etc.
- earthmoving*
- working at height*
- work in risers or lift shafts*
- working over water*
- working with electricity or gas services*
- using mobile elevating platforms*
- scaffolding*
- working with hazardous substances*
- vehicle movements*

*Delete as applicable

Site set-up

Welfare facilities

- number and location of WCs
- number and location of urinals
- number and location of wash hand-basins
- how will washing facilities be provided?
- how will drinking water be provided?
- location of drying room facilities
- location of mess room facilities
- location of canteen facilities (if applicable)
- are toilet/welfare facilities anticipated to move during the project? If so, detail new location and alternative possibilities
- detailed site plan to be included.

First aid

Names of trained first aiders/appointed persons

Copy of certificates to be attached to this document

Location of first aid kits

Health and safety cabinet

Location of first aid room (if applicable)

Incorporate into welfare facilities

Accident/incident reporting

How will accidents/incidents be notified and by whom?

What records/systems will be kept on site?

Who will investigate site accidents/incidents?

How will site personnel be made aware of site safety issues raised by accident reporting?

Will a disciplinary schedule be kept on site?

Disciplinary records will be kept on site, subject to action taken

Facilities for clients and site visitors

Provision of toilet facilities — what and where (including female facilities)

Meeting room

Signing-in procedure

Site safety meetings

How often will safety co-ordination meetings be held and who will attend?

Co-ordination of a multi-occupied site

Will the site be multi-occupied? YES/NO

If so, how will the sites be co-ordinated?

Site access, storage and waste disposal arrangements

Pedestrian operatives/visitors access

Vehicle access

Delivery access

Location of and route to site office

Details and location of site storage facilities

Site waste disposal facilities

Training

What induction training will be given?

Who will conduct induction training?

Who will receive induction training?

Where will it take place?

Where will the training records be kept and by whom?

How often will induction and other training take place?

How will contractors' and sub-contractors' safety training be checked?

What training and communication in respect of site risk assessments will be undertaken and by whom?

All tasks are to be carried out by competency persons and recommendations of the risk assessment adhered to

Statutory inspections of equipment

What equipment will be inspected, how often and by whom?

Are scaffolding inspections envisaged? If so, who will undertake the inspections and are they fully trained? Certificate to be included

Where will records be kept?

COSHH

What hazardous chemicals are likely to be used on site?

Who will carry out COSHH Assessments?

How will COSHH Assessments be checked?

In what format will they be?

How will use of substances, generation of dust, etc. be controlled on site?

What PPE will be provided for dealing with harmful substances?

Control of noise

When will noise assessments be carried out and by whom?

How will noise be controlled, i.e. at source, PPE?

Monitoring and auditing site safety

Who will review health and safety on site, and how often?

What procedures will be adopted?

Who will audit records?

Site rules

What site rules are envisaged to deal with the following (the list is not exhaustive)?

- Identify what activities are relevant to this site
- Working at heights
- Excavations
- Demolitions
- Erecting scaffolding, etc.
- Using hoists, lifts or cranes
- Maintaining 110 v supply
- Using portable electrical equipment
- Manual handling
- Using welding equipment
- Hot Works permits
- Permit to Work systems
- Working with polluted water
- Rodent/insect/pigeon infestations
- Temporary support works, formwork
- Roof works
- Removal/installation of glazing
- Provision and use of work equipment
- Removal of waste
- Use of hazardous substances
- Controlling dust emission
- Controlling noise emission
- Testing and commissioning equipment

Please list any specific activities as detailed in the Pre-tender Health and Safety Plan.

Health and safety file

Who will be responsible for collating information for the health and safety file and forwarding it to the Planning Supervisor?

Appendix G

Contractor competency questionnaire

Regulations 8 and 9 of the Construction (Design and Management) Regulations 1994 (CDM) require appointed contractors to be 'competent' and adequately resourced with health and safety. The following information is therefore required and should be completed as fully as possible.

Please add additional pages as and where necessary, stating the question number you are answering.

About your organisation

1. Name and address of organisation:

   _____

   _____

   _____

   Postcode: _____ Tel no.: _____

2. Nature of organisation:

3.   Name of head/senior person responsible for overseeing health and safety within your organisation (please print):

What health and safety qualifications does this individual hold?

Is the individual named solely responsible for health and safety or do they hold other positions within the organisation? Please state.

If the position is multi-functional, how much time do they allocate to health and safety issues per month?

4.   Please list professional bodies of which your company is a member/accredited.

Are you quality assured following an accredited quality system? If so, state which system and attach certification of accreditation.

Managing safety

5.   Do you have a health and safety policy?   YES/NO
If yes, please attach a copy.

How do you emphasise management commitment to health and safety within your organisation?

6. Have you ever been subject to any statutory action in relation to health and safety as an organisation or has any individual within the organisation: including improvement notices, prohibition notices, prosecutions or cautions and, also, civil claims? If the answer is 'yes', please provide details of the action taken, and measures taken to prevent a recurrence.

7. Have you had any RIDDOR incidents within the past five years? If the answer is 'yes', please provide details of the incidents and measures taken.

8. Who is the 'competent' person(s) within your organisation with regard to the Management of Health and Safety Regulations 1999?

9. Do you rely on in-house or external expertise with respect to health and safety?

   If external, please give name, address and telephone number of external organisation.

   If internal, please give name and position within the organisation and provide a CV of the individual named.

   How often do you have access to this advice?

10. How do you obtain information in respect of health and safety to ensure that your organisation is kept up to date in respect of new legislation?

11. Please give details of active and pro-active health and safety monitoring undertaken by your company, or on your behalf. How regularly are these undertaken?

12. Do you act as a Principal Contractor?

    If so, how would you manage health and safety in a multi-occupied site?

Risk Assessments and method statements

13. What procedures do you have for completing Risk Assessments?

    What training do operatives have to ensure they understand the need for site-specific Risk Assessments?

    Who compiles method statements and how do you ensure that these are communicated to the workforce?

What procedures do you adopt for training operatives on method statements and for ensuring that they understand their importance?

Managing sub-contractors

14. Do you sub-contract any activities?

15. How do you assess that a sub-contractor is competent in respect of health and safety?

Do you use sub-contractor competency questionnaires?

YES/NO

If so, please attach a copy.

Do you limit the number of sub-contractors which you will have on a site at any one time, if so, what limit do you set?

How do you sign off sub-contracted work on completion?

16. How do you ensure that sub-contractors work safely on site and what action would be taken following an unsafe system of work by a sub-contractor?

Maintenance of equipment, plant, etc.

17. Is the equipment you use on site primarily your own or hired in?

How do you assess the safety of electrical equipment used on site in respect of electrical safety and appropriate guarding?

Occupational health

18. Do you have a formal procedure for completing COSHH Assessments? If so please attach a copy.

Who undertakes COSHH Assessments and how are these checked within your organisation? What qualification do they hold or what training have they received?

What systems do you have in place to try to eliminate the use of hazardous substances?

19. How do you eliminate or minimise manual handling activities on site?

What training do your operatives receive on manual handling? How do you convey information on manual handling to sub-contractors?

Who is responsible for completing manual handling risk assessments and how often are they done? (Please supply examples.)

20. What approach do you have to noise control on site?

21. What procedures do you adopt when asbestos is either known to be on site, or is found unexpectedly?

Assessment and management of resources

22. Please provide details of resources available to you as a contractor. Include number of staff available for professional input, administration, etc.

23. What is the maximum number of projects that you are able to work on at any one time? Please be as specific as possible.

24. What type of work do you consider to be your speciality?

25. Please give examples of similar works you have undertaken, including names of Clients who can be contacted for references.

26. Are there any restrictions in respect of work you are able to undertake, e.g. distance from head office?

Training and competency

27. What training do site agents and contracts managers receive in respect of health and safety?

28. How do you ensure that sub-contract employees and visitors on site are given specific information concerning that site?

29. What level of training is given to site operatives?

    How many staff have received first aid training and which organisation did they receive their training through?

Construction phase health and safety plan

30. What procedures do you have for ensuring that this is site-specific and relevant to the project in hand, e.g. who completes it?

The following documentation must accompany the questionnaire:
31. Your Company Health and Safety Policy.
32. Evidence of safety management procedures, including safety auditing procedures.

33. Example method statements and Risk Assessments from a contract you have undertaken in the past twelve months.
34. Example COSHH and manual handling assessments from a contract you have undertaken in the past twelve months.
35. All information on any criminal charges in relation to health and safety must be provided. This should include details of improvement or prohibition notices, prosecutions or cautions.
36. Information on any civil claims under Employer or Public Liability Insurance, or any other civil action claims.
37. Evidence of liability insurance must be provided.

Please provide names of two Clients or planning supervisors (including telephone numbers) outside of your organisation, who can act as referees and vouch for your competency in respect of health and safety.

Name of company _____

Name of contact and position in company _____

_____

Telephone no. _____

Name of company _____

Name of contact and position in company _____

_____

Telephone no. _____

Finance/insurance details

i) Please attach copies of your current Public and Employers Insurance Certificate.
ii) Please attach details/copy of your CIS Tax Deduction Certificate.

Appendix H

Checklist for approving contractor competency

Checklist on following five pages.

Checklist for approving contractor competency

Score 2 = satisfactory
Score 1 = minor information missing or additional information required
Score 0 = unsatisfactory response or information not provided

Contractor:

Date information received:

Contractor deemed competent?   YES/NO

Percentage score:

	Score	Comment
1. Name and address of organisation		
2. Nature of organisation		
3. Head/senior person responsible		
4. Membership of professional bodies		
5. Contractor has a health and safety policy?		
6. Legal proceedings		

7. RIDDOR incidents

8. Who is the 'competent' person(s) within your organisation?

9. Expert safety professional available?

10. Keeping informed

11. Active monitoring

12. Management of health and safety on a multi-occupied site

13. Procedures for completing Risk Assessments/method statements

14. Sub-contract activities

15. Assessment of sub-contractors.

16. Sub-contractors work safely on site?

17. Equipment on site primarily owned or hired in?

18. COSHH Assessments/hazardous substances

	Score	Comment
19. Manual handling activities on site/manual handling training/Risk Assessments		
20. Noise control on site		
21. Procedures for asbestos on site, or if found unexpectedly		
22. Resources available as a contractor		
23. Maximum number of projects		
24. Type of work considered to be speciality		
25. Examples of similar works previously undertaken		
26. Restrictions in respect of work able to undertake		
27. Training for Site Agents and Contracts Managers in respect of health and safety		
28. Sub-contract employees and visitors on site are given specific information concerning that site?		

29. Level of training for site operatives/first aid training/organisation	
30. Procedures for ensuring that Health and Safety Plan is site-specific and relevant to project	
31. See Item No. 6.	
32. Evidence of safety management procedures, including safety auditing procedures	
33. Method statements/Risk Assessments from a contract undertaken in the past twelve months	
34. COSHH and manual handling assessments from a contract undertaken in the past twelve months	
35. Information on any criminal charges in relation to health and safety. (Including prohibition notices, prosecutions or cautions.)	
36. Information on any civil claims under Employer or Public Liability Insurance, or any other civil action claims	
37. Liability insurance	

Penalty marks (Lose 10 marks for each item)

Absence of/insufficient health and safety policy

Serious legal proceedings

Poor documentation (general)

Risk Assessments not suitable/sufficient

Absence of qualified competent persons

Lack of training reflected in health and safety

Total score available	74	
Score achieved		
Penalty points deducted		
Overall score		

Percentage		
Target	75%	
Variance		

Did you contact referees   YES/NO

Was response favourable?   YES/NO (attach further information as relevant)

Assessed by: Date:

Appendix I

COSHH Assessment Form

CONTROL OF SUBSTANCES HAZARDOUS TO HEALTH REGULATIONS 2002

Company:

Address:

Contact:

Product:

Job task:

Application:

Equipment:

Safety data sheet attached: YES/NO

RISK IDENTIFICATION

Hazardous component(s):

Hazardous nature of component(s):

Health hazards (known):

Persons affected:

Duration of exposure:

Level of exposure:

RISK CATEGORY

Control measures

For users:

For persons in location:

Storage:

Training:

Health surveillance:

Re-assessment:

Date of assessment/revision: